LIVING GREEN

LIVING GREEN

A Summer's Cycle around Green Britain

CHARLES HOULT

GREEN BOOKS

First published in 1991 by
Green Books
Ford House, Hartland
Bideford, Devon EX39 6EE

Cartoons by
Axel Scheffler

Typeset in Bookman by
Chris Fayers, Soldon
Devon EX22 7PF

Imageset by P&M
Exeter

Printed by Biddles Ltd
Guildford, Surrey

British Library Cataloguing in Publication Data

Hoult, Charles
Living green.
1. Great Britain. Ecology
I. Title.
574.50941
ISBN 1-870098-40-4

Contents

I'm a little tired of this 'what's wrong with the world' thing. We've got a pretty good idea what the problem is. It's mostly in people's heads.
Andy Moore

The optimists believe nothing needs to be done, the pessimists believe nothing can be done.
Nigel Wild

So often ideas are basically simple but you need to bring together several to lead to action.
Roland Chaplain

Alternative Technology is about making it feel different to be alive.
Peter Harper

If you live by habit and not by question, you are making no progress
Brig Oubridge

I think people know that there are cataclysmic times ahead and that changes are creeping over us. People dream freely of an alternative. They are preparing for this in their subconscious.
John Lane

It seems obvious to me that the change needs to happen in a radical way—and radical means deep, revolutionary, far-reaching.
John Button

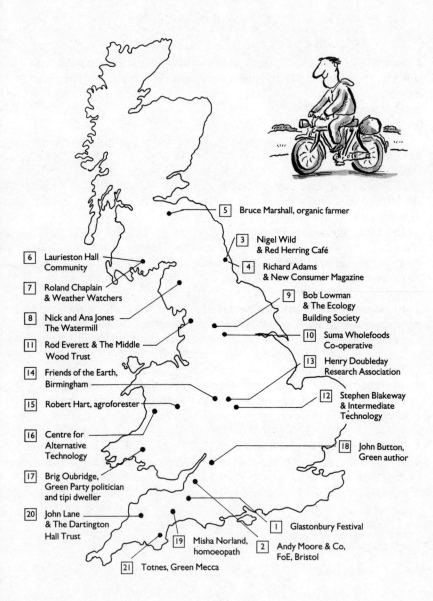

5 Bruce Marshall, organic farmer

3 Nigel Wild
   & Red Herring Café

6 Laurieston Hall
   Community

4 Richard Adams
   & New Consumer Magazine

7 Roland Chaplain
   & Weather Watchers

9 Bob Lowman
   & The Ecology
   Building Society

8 Nick and Ana Jones
   The Watermill

10 Suma Wholefoods
   Co-operative

11 Rod Everett & The Middle
   Wood Trust

13 Henry Doubleday
   Research Association

14 Friends of the Earth,
   Birmingham

12 Stephen Blakeway
   & Intermediate
   Technology

15 Robert Hart, agroforester

16 Centre for
   Alternative
   Technology

18 John Button,
   Green author

17 Brig Oubridge,
   Green Party politician
   and tipi dweller

20 John Lane
   & The Dartington
   Hall Trust

1 Glastonbury Festival

19 Misha Norland,
   homoeopath

2 Andy Moore & Co,
   FoE, Bristol

21 Totnes, Green Mecca

Introduction

I drank too much coffee that evening, and lay awake in my squalid bedsit. Half my mind wrestled with my ambition to succeed at journalism school, half was haunted by thoughts of a world in turmoil. The old ways had not proved the best ways. Ecology, economy, society, all looked tarnished. And they rocked the Berlin Wall as I learnt shorthand.

I saw my way out: write a book. It would be simple, practical and optimistic. I scribbled an outline and went to sleep. Next day I sent it to Satish Kumar, editor of the magazine *Resurgence*, with some added padding about Cardiff Centre for Journalism Studies being Britain's best.

Dear Satish Kumar,

I have a good idea for a readable book. I want to write the diary of an odyssey or pilgrimage around Britain in search of people who live by ecological principles. The book could help to define some of the facets of this new culture by showing people's idealism, practicality, even ennui. The idea is to present people, to people, in order to show possibilities. There will be thinking, objections and visions as well as a journey from place to place.

I would like to lay minimum plans and follow the suggestions and leads offered by people on the way—organic farmers, bakers, writers, recyclers, craftsmen and shop-keepers.

I will travel, with ecology in mind, by bicycle.

New Greens, myself included, clamour for ways to act to avert ecological disaster, but so many lack realistic (real) role models. A book like this could offer a spectrum of living examples with their own patterns of conflict and compromise.

What do you think?

Yours etc etc.

To my surprise John Elford, of *Resurgence's* sister company Green Books, replied quickly with an ominously thin letter. All I remember he said was 'excellent'.

We met in early December 1989 when Satish came to lecture in Bristol. I drove, coy that they would mark car ownership and carbon dioxide against my Green credentials. The rendezvous was Greenleaf Bookshop in Colston Street. A small bearded Indian gentleman stood by the counter with a funny woolly hat over his ears. He kindly told the shop assistant he ought to display Green Books more prominently. Behind him waited John, a taller man with Satish's notes tucked under his arm like a solicitor in court.

They were totally charming when I introduced myself. I felt they were initiating me into a magic ring of dedicated people, an impression accentuated by the dark cloak of a brisk winter evening and the smell of Christmas down steep alleys. Satish bought tea and sized me up. Here was a man wholly Green in walk and talk, who, as a monk, had journeyed penniless from India through the western world in search of peace; there was I utterly fresh to it all.

Satish was effusive with his ideas and enthusiasm so I did not have to stick my neck out on the minor points of detail. He and John somehow liked me and ordered me off for a weekend to see Dorothy and Walter Schwarz. Dot would supervise my work. Her husband's respected gaze would be distant but vital as a seal of approval.

Two months later, I arrived to vodka with the Schwarz family and Andy Langford, another aspirant author, of the permaculture persuasion. Walter racked our brains for ideas he might incorporate in a Channel Four series. I felt queasy when it was my turn for suggestions, and changed the subject to *The New Dissenters*, a book Walter had written as a result of his work for the *Guardian*. I was not used to discussing the ideas of the day with those who forged them. *The New Dissenters* had captured my mood and crystallised a new nonconformist conscience emerging in Thatcherite Britain.

'There's a different set of values running along in the country parallel to Thatcher which actually hasn't got a voice to express itself at the moment,' said Monsignor Bruce Kent, CND chairman and Catholic priest, to Walter in the first chapter in his book. And Walter hinted that the Green movement was a home for these dissenters, intellectually speaking.

Over lunch he explained. He had seen it all before as *Guardian* correspondent in Europe, when he covered the rise of the German Green Party. He had even joined their procession to the Bundestag when they took their first seats in parliament. Walter's recollection brought the occasion alive: the atmosphere of optimism, the flowers, the radical outfits, the celebration. I went away determined to have such good stories to tell.

The hardest thing has been to breathe life into the page, but that was what I set out to do. I hope you find this book an antidote to nightmarish news and grey statistics, to contorted intellectual postures and to dunces' guides. I hope there is some balance.

This is a travel book with a strong theme; it is a book of stories; but most of all it is a book about simple ideals and values that can give a fresh perspective to the complicated lives we lead in Britain in the 1990s. If we could all live like the communards, artisans and enterprising people I have met, the world would be definitely a safer, Greener, more satisfying and more exciting place.

CHAPTER ONE

Glastonbury Festival

I stood between an old oak hung with yellow ribbons and a
rockery maze. The valley below swelled with thousands of
tents for a hundred thousand people, tipi circles, bright coloured
flags, music. I gazed across Glastonbury Festival, the best place
to start my summer's journey.

I had come in search of Britain's Green people. Friends joked
about martians. I wanted to root out as rare a breed: people who
actually lived and worked by ecological principles. Fine words are
easy: but how do the really committed live? What conflicts and
compromises dictate their lifestyles? Could I learn from their
example?

I planned a cycle tour to find interesting characters, hoping they
would tell their story in exchange for a contribution to their work.
I could then pass on their perspectives as I went. A pilgrim, a
questor, a story-teller. I had visions of Chaucer and Sir Gawain
and the Green Knight, medieval adventures with knights and
white chargers, as I looked out over the mass of pilgrims to this
festival. The gathering struck a very basic human chord of
celebration, though in homage to the modern bards of rock and
rap.

It began to rain, a daunting prospect for a homeless city boy. I
had finally left Cardiff equipped with journalist's shorthand, via
City Hall where I signed off the poll tax register. Reason: cycling

around the country for three months. Forwarding address: none. Charles Hoult, English graduate, 23, of no fixed abode, one key and a NatWest bank account. My key, I admit with remorse, belonged to a car at this stage. I drove down with two hitchhikers and used it as a caravan until I joined friends with a tent. I cheated on my cycling commitment out of convenience, out of laziness and to cover the miles before I returned home to Newcastle-upon-Tyne. From there my pedals would begin to turn and I could park my conscience and garage the car.

I was ill-prepared for rain at the festival, on 200 acres of Somerset pasture, living in a tent, with no loo and a tiny store of dried food. Miserable. So, high up in the Green Field with views of Glastonbury Tor obscured by the shower's mist, I sheltered in a tipi, mobile home to a gruff Scotsman. We sat on animal pelts as the shower passed, beside a fire expertly dug into a pit. The spacious home, worth £650, was light and free of smoke despite the wind and gloom outside.

My temporary host lived in Wales but had pitched on the festival's Green Field among healers, travellers, campaigners and traders. The atmosphere had elements of a college freshers' week: wise men and women at stalls, on soapboxes, touted mind-expanding initiatives for wind power or for tribes endangered in far off lands. The blacksmiths and beer tents and hawkers gave the impression of a Thomas Hardy country fair. The hippie travellers' anarcho-punk look, their conformity from army surplus stores, was more out of the film Mad Max. They loitered on track corners and muttered, 'Hash for cash...whizz... ice... World Cup acid... E... trips here... hash and honey truffles.' The police roamed and looked dead ahead. Everyone else was on the march, determined to enjoy the escape from brick homes, relations, cares and cars.

I was persuaded to come to Glastonbury because the Green Party's newspaper offered free tickets for the festival if you were prepared to work for six clear-up shifts to pick through and recycle the rubbish and donate your wages to the party. They needed the money and I did not have to join the party so it seemed like a good idea. I went to sign in for early morning rounds.

The introductory meeting was fraught. Somebody had laid on a team to coerce the punters not to chuck litter into the mud in the first place and christened them the Green Police. Shivers went down the soggy marquee. Greens hate authority and they hate hierarchy. Everyone else had non-confrontational titles, but this did not make up for the gaffe. Team leaders, called co-ordinators, did not chair but facilitated the meeting. The uproar subsided, only for the meeting to drone on over volunteers' diets. Free food was vegan, but what special preparations were made for gluten-free?

I found the explanation for the muddled behaviour and the talk of 'appropriate structures' in a copy of *Green Line* magazine. '"Anarchic decentralists",' wrote Steve Dawe, 'are the nearest thing the Party has to a real faction, with supporters on the key organisational committees of the Party... they express their views through the "Other Ways of Working" Working Group.'

The faction's supporters challenged the conventional political process in the name of more user-friendly discussion and greater participation in decision-making by grass-roots members. My observation, and Steve Dawe's conclusion in the article 'Are we in fact gazing at the same navel?' was that 'In the name of empowering people, they waste a lot of time.'

The morning after the heady, drizzly, smoky night before, we met our team again and went forth to rustle beer cans from the mud by the main Pyramid stage. Janet Alty, my team organiser, bounced about like a hockey coach, dressed in a sweatshirt and fuelled by a good night's sleep in her well-equipped and well-erected tent. I teetered after five hours' sleep in the boot of my Vauxhall Nova, just too short to let me stretch outright. Janet had few qualms about scolding late risers and giving them extra detention.

She told us, as she doled out our breakfast tickets, that she volunteered sixteen hours a day for the Greens and had personally co-ordinated their economic policy review. She said she would have classified our labour as part of the 'gift economy'. It was a concept she wanted built into the Gross National Product so that the economy valued goods and work exchanged without money.

The thought filled me with dread as I bent double in front of the passers-by who thought I had been paid to clear up their mess. How could this expansion of bureaucracy work in practice when it was hard enough motivating volunteers to recycle over a million cans for the three day event?

By Saturday, I had settled into a routine of minimum sleep. Everything happened in the evenings, when a dark hush fell on the tented fields and the revellers trooped between jazz and theatre and circus with flaming torches and pockets full of change from buying beers and trinkets. The weather cleared for the day and, between snoozes in the sun, I had the opportunity to meet several people I felt would fit the bill for my tour. I bumped into Andy Moore, southern region projects co-ordinator for Friends of the Earth, and discussed various ideas. We had first met at the launch of Cardiff Recycling City a month earlier, after I had managed to corner Dave Gee, the FoE Director, for a brief chat about his trade union background.

Andy was delighted to sit back and ponder *omphalopsychics*, the navel-gazing tradition born of ancient Greece. This sort of stuff had made the great Glastonburys, he told me, and the same feeling had been reinstated in the Green Field as a conscious effort on the part of the organisers to make 1990 more than a drugs and rock festival. There we were, amid pyramids and tents fit for jousting knights and windmills, when the Women's Environmental Network rolled up in their camper van. They were most miffed to find the women's camping area had been invaded by men and Friends of the Earth helpers were flustered when the van was parked such that it obscured their stall.

Andy kindly whispered directions to the Real Meat Company's food stall where they sold tasty organic meat, whatever that meant, and I slipped away from the fray. The Green Party had contracted the Leicester Works Brigade for Nicaragua to feed litter pickers and they had gone for a complete vegan menu. I had a pang for bacon butties.

I tramped across the valley for my nosh. Hari Krishna devotees and Christians handed out leaflets for free meals and recycled personalities. Free enterprise included matches and cigarettes, Moroccan lanterns and a last minute delivery of army surplus wellies. A Rastafarian stall pulled more crowds than most because the owner had a television set on which to watch the Argentina World Cup match. Above, a huge crane dangled a cage of lunatics who periodically jumped out with elastic tied to their ankles.

I satisfied my hunger, then circled the fields of orange, green and blue tents with groups of friends pitched in circles, each with their own bonfire of scavenged or chopped down wood. I returned to Friends of the Earth. The Women's van still blocked the view.

Andy introduced me to Oliver Tickell, son of Britain's UN Ambassador Sir Crispin Tickell, the man reputed to have focused Mrs Thatcher's mind on the global warming link with the tropical rainforests. Oliver lived in Oxford where, apart from holding a pivotal role in the *Green Line* editorial collective, he had set up Earth Warriors, a campaign group fighting for the heritage of indigenous tribes. He peered down through his spectacles in slightly donnish manner, and chatted about his particular corner of the Green movement. He had once ventured to Burma in a plan to save the rainforest there, but backed down when he realised that he could not better the locals' work. His campaigning subsequently centred on support for local-based initiatives which he could help directly through personal contact.

4

Martin Fodor was also busy on the FoE stall. He worked at the Bristol FoE's own warehouse, the Avon Environmental Centre, on a scheme which recycled all sorts of scraps of rubbish for children to use in playgroups. He recommended I speak to Godfrey Boyle on my tour. Godfrey had been involved in a very influential magazine called *Undercurrents* and lectured for the Open University Energy and Environment Unit despite the fact that he had no degree himself.

I browsed through the other stalls, and popped in on the Green speakers' forum where debate raged on Green consumerism, land reform and women's rights. The 'Other Ways of Working' Working Group had obviously got to the tent before me, because a novel system had been adopted to run the discussions democratically. Anyone was allowed to participate, but they had to wait to be handed the 'baton', in this case a shoe, by the previous speaker. As a speaker, you could choose whom to nominate next—much to the frustration of the billed guest speaker, whose paper was being berated from several fronts when I nosed in.

Across the gravel path I spoke to Kate Sebag on the Tools for Self-Reliance stand. She explained how volunteers and a small special-needs staff in Britain collected and repaired old tools for despatch to the developing world. Communities could then literally build their own lives with their own hands.

Paul Harrison at the Ethical Consumer Research Association caravan sold me an edition of his magazine which dealt with the vested interests of the bicycle industry. I would now be better informed about who would benefit from my punctures in the Pentland Hills, I thought.

Later, I spent some time on the Permaculture stand, talking to Andy Langford, one of the movement's 'designers', whom I had met on a visit to my publisher's commissioning editor, Dorothy Schwarz.

Permaculture grew up in Australia around a guru-like figure called Bill Mollison who decided that the best way to save the world from ecological disaster was to replenish the land around us with sustainable small-scale agriculture. He researched the most effective ways to plant his plot to make use of advantages seen in the wild—diversity, hardy varieties, perennial instead of annual crops. The whole thing spiralled into a complete and radical way of life that cut down on needless consumption and energy where an alternative could be found.

I straddled a hay bale, chatted and listened, rather like a disciple or a student of Pythagoras, as I followed the experts' talk. An ordinary-looking bloke came up to debate the most effective design for a methane generator attached to a compost heap digester. Gorse bushes came under scrutiny as good soil-builders. One elegant, elderly woman bought a copy of every single book on display (some sort of socio-economic indicator since the *Permaculture Manual* alone cost £40). She picked up on Andy Langford's idea for better information with talk of a radio station. Andy, it turned out, had already looked into the idea and got quotes for buying pirate broadcast equipment. I never dreamed things had gone so far in the struggle to give people ideas about planting their own patch. What Andy felt was needed was a network which could share the ideas pioneered across the gardens and farms of Britain: successful small-scale technology, wild plants suited to the temperate climate, tips on how worms could revamp your soil.

The talk was of soil. I had already got a lot closer to it, scrabbling about with a pair of scavenged gardening gloves and a little stick as my tool for prising out the most embedded ring-pull or crisp packet. I was infuriated by the rubbish left by the thousands of trippers out for a good time without a care, divorced by city living from the real ends and means of waste collection and disposal. Oh, city dwellers, what have we forgotten?

Soil came into the conversations at supper; so did squats and social work and marijuana and the anguish of transferring to menial, hard labour for three days. We shared plates and cutlery, because these were above budget. I borrowed a set from a life model who posed for students at Brighton Art College. He was fawning with friendliness and showed me his tattooed chest and nipple rings. Draco, as he liked to be called, had wonderful style. He had paraded in an orange boiler suit with his head totally shaved and topped by a bright Indian skull cap. He beckoned me near to describe his days in the Paris of the sixties. He had performed for Andy Warhol in a film, he had squandered everything on the best couture. He still had his figure at the age of nearly fifty and had just had a Prince Albert ring slipped on the end of his penis. I did not ask to see it, but invited him to come and watch saxophonist Courtney Pine on the World Music stage.

It was Monday. The final songs had been sung and all the drug pushers arrested when they left the site, because the police had photographed them from Land Rovers. The litter pickers massed for an all-out sweep of the top fields. I joined Julie Mountford and Sally Davis, new-found friends from Liverpool Green Party (never forget that each area has its own autonomous party, divided into branches and even 'twigs' with the power to decide its own policy regardless of national ideas). We were a maverick team and moaned that 'jolly hockey sticks' Janet had been replaced by a lazy co-ordinator who did not pull his (substantial) weight. We linked up with a friendly looking scouser called Simon O'Brien, who, it turned out, was an actor from Channel Four's *Brookside* soap opera. He curried no special favours from the party and was as grubby as the rest of us, even though he had spent a few shifts behind the scenes for BSB's *Power Channel*.

By supper we had filled sacks and sacks of rubbish across the acres; blue to recycle, black to bin. I was filthy. I had washed in a teacup for three days but needed to scrub up before I landed on Andy Moore to tackle Bristol. In a field amid the healers' tipis there was rumoured to be a sauna with a shower. Julie and Sally went in search of a wash-house where Green Party members had priority in the queue.

I went off and discovered a Heath Robinson contraption beside the hedge, a shed made from a huge oak barrel with a fire ablaze in a cut-out space. Bits of turf and orange plastic sheet stopped the gaps. Outside was the shower, a hose tied to a branch— without so much as a leaf between me and the madding and milling crowd. I felt very Anglo-Saxon as I shivered and looked at this wholly Scandinavian experience.

For youth, hygiene, Glastonbury and Green Books I was there under the tap, and it was only a small step to follow the naked men and women into the sauna.

Inside the pitch dark, blasted heat refreshed and blamed me. Stiff backs wilted with another splash of water and herb blend on the coals. A spontaneous round of chanting swelled up as we breathed and basted. Our eyes adjusted to the dim. I was in the pine forests of Finland.

A nubile newcomer fumbled her way through the inner door, obviously still cloaked in self-consciousness. 'You are the only one in the room who's naked,' an American wag suggested. Our laughs were a huge release of embarrassment. We were having fun. What better motive for my trip: fun from shared experience, the chance to show others they have common ground and fellow travellers on the tricky journey toward change.

My last visit in the Green Field was to the Croissant Neuf circus tent whose technicians from the Centre for Alternative Technology, Machynlleth, had harnessed the wind to power the show, weather permitting. Dilwyn Jenkins explained their aim of upgrading the centre through a share issue worth £1 million.

'Will it be a good investment?' I asked.

'What do you mean by good?' he replied, leaving me speechless. I was learning a new set of values.

Then I overheard one long-lost friend ask another, as if it were a novelty: 'Do you live in a house, now?' I had entered another world.

The word was that the trouble began at one of the festival gates. A fight flared between on-site security staff and travellers over an old grievance. Someone threw a petrol bomb. A wall of police appeared from nowhere and the travellers scattered.

I was basking in the afterglow of my sauna when news of the riot spread to the Greens' corner. We heard the roar of the mob, but assumed it was further festivity. The travelling folk were angered at their rough treatment at the hands of the private security people hired by Michael Eavis, who ran the festival on his farm land. Usually they had jobs for a week to clear up but the Green Party volunteers had taken them. They had to move on more quickly than usual. We, who had replaced them, were trapped with the travellers' field between us and the open road.

I went down to see what was going on. In the emptied field where I had caught a snippet of the World Cup on television, fifteen punks half-heartedly jabbed at a stolen Land Rover, with more noise than thunder. They rolled it over, to shouts from provocateurs at a safe distance, and stacked cardboard under the fuel tank. 'Go on! Burn it!'

A lone voice screeched on a megaphone: 'It's stupid, stop, it's stupid. This isn't Eavis's. It isn't security's. It's just the plumber's. It's stupid.' The voice of dissent was drowned by derision. But a second speeding Land Rover dampened the fray. Was it police? Security? It was the mob's own and it screeched up and saluted victory from where the back door had been. Adrenalin burst through again.

'You know what to do,' the driver snorted.

I hated the anger in his voice. It was cool and premeditated. He knew shared blame was best. They smashed windows with the iron marquee pegs and a goon dived inside, only to slide out again as they rolled and 'torched' the vehicle with petrol bombs. Everything slowed down for me on the sidelines. The bang, as the petrol tank exploded, was just a whimpering phut.

I felt heavy in heart, my blood curdled, as I turned away. I found no reason for the mindless violence nor did I understand the compulsive urge I had to witness it. We walked back to camp as others, with a different reaction, sped to the fight, equipped with billiard cues and axe handles whisked from nowhere.

At camp, the disco was hushed and the earlier talk of non-confrontation on the political platform had turned into an opposition of blood and knives and drug cocktails. Men were men in this situation. The DJ mustered a 'keep calm' and periodically reported on the bush telegraph's latest news. I told him I had seen a gang of partisan troops armed to the teeth, huddled on the back of a truck and bound for rampage. It all recalled times when loins were girded and a single topic was on all lips. And the flashing lights danced on the canvas marquee. The Titanic. The War. Stiff upper lip. In a crisis, the intellectual responses to conduct become instinctive ones of self-preservation.

A watch was posted consisting of the biggest men—to monitor and not to confront. They did not pick a wimp like me, but I joined in for a few hours of 'halt who goes there', then took a stroll into the dark zone at three o'clock in the morning. The information office was ablaze. I stood where I had done earlier, mesmerized by Courtney Pine, alone but for the bright fire.

A police helicopter banked only seventy feet overhead, close enough to ruffle my hair. It hovered and I froze. I was alone in the glare of its searchlights, just me and the grass in the focus of telephoto lens or telescopic sight. I did not dare to move. The power of the state held me in its stare—and I was frightened.

CHAPTER TWO

Andy Moore, Matt Dunwell and Pat Fleming, FoE Bristol

The melodrama subsided with the dawn. I drove to Bristol dog-tired. Over a cooked café breakfast I read *The Independent* to recover the mundane values of everyday. None of the night's ravages was reported.

Andy Moore, Friends of the Earth's southern region projects manager, with an eye for real meat, put me on an even keel as we shared some beer in Montpelier that night. He kindly laid on all available contacts—another at the table was Rob Tomlinson, manager of Western Eye Television, who gave TV and media training to the top Green Party members. Andy racked his brains to coin an apt Green phrase to write into my book for posterity.

He introduced me to 'eco-lala' which turned up in Jonathon Porritt's *Coming of the Greens*. The American social ecologist Murray Bookchin had first coined the term to rubbish deep Green concepts which had grown up around American Indian philosophy and post-industrialist paganism.

'Murray,' said Andy, 'was not having any truck with this borrowing the earth from our children stuff which people trot out at all and any occasion, and which managing directors of rather hungry capitalist organisations spout to each other on their promotion videos. He didn't believe there was a higher race of warriors. He called this eco-lala.'

Andy was more influenced by the accessible. He advised me to start my intellectual odyssey with *Far from Paradise* by John

Seymour and Herbie Girardet and *To Have and to Be* by Erich Fromm, both antidotes to 'Indian mystics sticking their big toes up their noses or sitting on poles for a very long time'.

The big man supped his pint and sagaciously advised me to track down James Hillman's 'The Soul at Work', an article he wrote for *Resurgence* before the 1988 Schumacher lectures. Andy was so inspired by it that he went along to hear Hillman speak. Characteristically he was late and without a ticket but, recognised at the door, he was ushered in before he could produce his chequebook. They even stuffed lunch vouchers into his top pocket. 'I nipped out to do the shopping and came back. I had missed the lecture, but heard the article was better," he joked.

Rob Tomlinson took us to his combined home and video studio; as dependents on car culture, we all drove in separate cars. We phoned for curry take-out and discussed my puzzlement at the pull of the South-West for all things Green. Was it a combination of money and distance from London?

Rob was more intrigued to look into the cultural climate that allowed Prince Charles to wear his environmental credentials on his sleeve. Did everyone foster something of a deep urge? He had interviewed a hardened truck driver a few days before, and between takes the man had discovered wild comfrey in the hedge. He had picked it to brew a remedy for his sick daughter, which he knew of because of his amateur interest in herbal therapy.

In the Western Eye TV office, Rob showed us a video from his and Glastonbury's early days, when the festival was more of a family affair. Then he switched to the Russian edition of a tool-makers promotional film, a ploy designed to nudge us back to the pub.

Over a pint, Andy told me of Frederick Harrison, a *Guardian* 'Grassroots' writer who was travelling in Ireland with two cats, Cheesy and Pugwash. If I was going to warm the outside world to the insider's view of the Green movement, Harrison had an accessible style to start from, he suggested. There was too much precious polemic about elsewhere. Harrison would play his mute cats in hilarious encounters that probed the mysteries of life.

Andy was fine company. He pointed me to a comfy new futon bed to sleep on after more beer and then a cup of tea with friends from the pub. I went to bed with his pub philosopher's theories, surrounded by piles of books and Andy's flatmate's half-finished carpentry projects. The light from the street lamps was in my eyes and I rolled over when the street erupted in screams and yelps. England had scored a goal against Ireland in the World Cup.

Andy's philosophy went like this:

'One: Apathy is innate in British culture. This side of our psyche would thwart any integrated, communal approach to the disasters ahead, whereas in Switzerland and Germany, where Green ideas have most hold, citizens are prepared to work for the good of society. They even churn the municipal compost heap once a week in their turn.

'Two: There's a new theory about the mystery statues on Easter Island. The tribe had cut down and ruined the trees and died out—a marvellous analogy for life on the planet. Before we arrived there were forests. When we've gone there will be no trees but some very strange things around for people to find, which will leave them guessing what we did. And after a while they will realise that we weren't that intelligent. We invented all sorts of electronic gadgets which were obviously incredibly complicated but we were actually so stupid that we managed to starve ourselves to death.'

He strongly believed in living life to the full and felt that he should share the 'straight' world he had set out to change. We had freshly-baked croissants and *pain chocolat* for breakfast.

The nature of his job also meant his life and his work were inseparable. Half his time was spent on trains here and there or rushing around London in taxis, the rest at FoE meetings where everyone else came for evening relief from their worldly jobs.

'There's a lot of fast food in this business,' he said, 'and life's generally disjointed. It's a way of life which, as they say, "calls for something else", and by the end of the day you really are looking for something else. I sometimes roll off the train at Bristol Parkway at eight o'clock at night too knackered to go to the take-away, let alone cook something.'

But Andy bought a word processor and reduced the time he spends at Avon Environmental Centre, his FoE base outside their London HQ. He has become a confirmed home-worker.

'I like the continuity. You get things done and can nip out to the shops when there isn't a rush on. It all fits into a day's work. I feel rested by the end of it. I don't arrive home to see that everything needs doing, I've done it and it hasn't taken a great chunk of my time. I actually feel a lot more contented and often more inclined to carry on into the evening.'

I wanted to shadow Andy for a typical day. His work had to be differentiated from campaigning. Most of the latter took place in Underwood Street, FoE's headquarters, where they have seven specific targets: Air pollution; Countryside; Energy; Recycling;

Rainforests; Water and toxics; and Global warming.

Andy was in the ludicrous situation of being in control of all FoE-backed projects from Kent to Land's End, including London. Travel was imperative and frequent.

He wrote letters on his home word processor until mid-morning, while I went off to visit Matt Dunwell, whom I had met in the beer garden late the night before. The deal I had struck with him was that I would come for a breakfast interview while Andy did some work, and then give him and his girlfriend a lift to the station to catch the London train.

Andy had filled me in on Matt. He was only twenty-seven years old but he had already made a big impact with Green groups. He had his own business, he had set up his own charity, and he had just bought a farm to put permaculture ideas on water farming into practice.

Matt took me up to his spartan bedroom, empty but for a mattress on the floor and an enormous paper lampshade. Downstairs one of his flatmates tinkered on the piano and the music swept up the stairs. It lingers on the tape of our conversation.

If there was one thing Andy wanted me to ask Matt, it was how on earth he had the time to organise so much. FoE had sent Andy on two time management courses to improve his skills. Andy had been late for both.

'It's interesting,' Matt said, 'to ask campaigners where they get their energy. Perhaps it was a joyous experience or a real shock they had when they were fifteen. For me it was a combination of seeing what can be done in Australia capped with the scale and the waste of destruction in Sarawak.'

Matt had made a snap decision to go to Australia. He had stayed in the North for a year after graduating in agriculture and environmental science at Newcastle University, but had then realised that he was stagnating with only odd days' work for the British Trust for Conservation Volunteers. At least if he were in Adelaide doing nothing it would be an experience. He recalled:

'As soon as I landed in Australia I just changed gear into a frenetic work mode. I travelled with the theme of networking through environmental groups, seeing what issues they dealt with, what type of people; whether fairly middle class or off-beat rainforest activists. I landed on people saying: "I've just come from x, y and z. This is what they are doing, what are you doing?"'

'It was a quest, I just sponged up what was going on. But it was a two-way process. I found I brought energy into a group. Say you have a group who are Men of the Trees in Adelaide, they have done good stuff for three years. Then somebody crashes in from

13

England who has got loads of energy, wants to learn lots. You are actually giving them a heck of a kick, to explain and analyse.'

Matt talked with a quiet voice, but he crackled with inspiration and leant forward in urgency. It was the first time he had been interviewed about something so intimate as his own motivation.

'In Australia, I learnt, there is incredible empowerment. If you feel strongly about an issue you just get up and march out on to the streets, lie in front of a bulldozer. In England, it's very much: "Oh well, we can't do this, can't do that." You can't get support for direct action, and the press is very wary.

'There was the issue of the Franklin hydro-electric dam in Tasmania before I was there. A man called Bob Brown, who is now a Green MP, decided to stop the scheme to dam the last wild river in the area. One thing led to another and the organisers campaigned for about a year. David Bellamy got involved. About 1,500 activists went to jail. It brought down the state government. It arguably influenced the general election which brought in Bob Hawke on a conservation ticket. In other words, it was seen as one of the most successful direct actions ever.

'I visited Tasmania after the hydro campaign, when the Wilderness Society was campaigning against wood-chip logging, where 300ft trees, some over a thousand years old, are turned into chips the size of a thumb nail for cardboard. It was a much more complicated issue to deal with and quite hard to sell to people but the development of this, subsequently successful, campaign was a really strong influence on me.'

Some 20 per cent of Tasmanian forest was later designated world heritage land by the Green MPs who came to hold the balance of power in Tasmania's state government.

'When I came back to England, I stopped off in Sarawak, Bali and Java. I saw some of the logging there, which was just one of the most awful sights. It was good experience in that I just felt desolate after seeing what was going on—complete destruction of whole cultures and wildlife systems. From an aeroplane you could see the topsoil pour out to sea. It looked like the whole place was bleeding.'

It was the contrast between Sarawak and Australia, he mentioned earlier, which drove him. He returned to Britain in 1987 and needed a job, so he went along to the London Ecology Centre which did something similar to the Australian Wilderness Society.

In every major town in Australia he had seen Wilderness Society shops selling books, postcards, T-shirts, recycled glass products. In Sydney alone they turned over a million Australian dollars a year. The shops' strength was their focused theme. People knew if they bought there they supported a good cause.

Matt had wanted to set up his own shop but instead joined the Ecology Centre as a receptionist for four months. He found it a nightmare dealing with all the different small-scale suppliers. They were generally erratic and would often get involved in their own campaigns and cut production off completely. So in January 1988 he set up Worldly Goods to act as middleman, marketing and distributing a range of chosen items through charities' catalogues and the like. Under the Worldly Goods umbrella, as with the Wilderness Society, you could be sure of 'sound' credentials.

In its first year the business had a turnover of £29,000. This rose to £120,000 in the second and £150,000 was forecast for the third. The one man band had taken on another employee, the business was 'more finely tuned' and the hope was for a contract with a High Street chain to have a pocket of each of their stores subcontracted off for Worldly Goods.

'Would you call yourself a campaigner now?' I asked.

'Yes, I would, but I decided I didn't like the confrontation way. When I was at university I worked on a straw-burning report for Friends of the Earth with a guy called Chris Rose. (Burning was actually banned after 1990.) It was very high pressure, aggro stuff. I'm much happier getting alongside someone and chucking over comments.

'I do still help Survival International. But with the start of Worldly Goods I don't so much do direct campaigning. I support groups through the work I do.'

'Did you think starting a business was a more pragmatic approach?'

'All approaches have a place. But you can't make very much money out of writing letters to your MP,' Matt said, wryly.

'But I gather you just bought a farm?' I said. There was a long pause punctuated by sighs.

'Yes, it's a difficult one. I inherited some money and I would rather it was in my own controlled property than in the stock market. Having said that, I think it's dodgy having more than ten acres if you are doing appropriate—in other words, low energy output—agriculture. Otherwise you need to charge around on tractors. At the moment we have two tractors on the sixty acres, but my long term aim is to get other people involved in the land

and end up with less land and fewer tractors.

'It's quite dodgy to buy a farm and say it's for the world's gain, because a lot of people would say: "Stuff you, mate." But I think you can have money and use it responsibly—if it involves people, if it involves environmental practices. That's a better solution than money in the stock exchange and keeping the status quo.

'I take one of the foundations of the permaculture movement to be looking at what people base their security on. A lot base it on money in the bank. In a way, that's the one thing that's accelerating global destruction. So much of it has been invested in environmental colonialism by the banks funding massive misguided projects. There's an element of my trying to pull out of that.

'Also, I want to try out permaculture principles in the UK. There is nobody using a good aquaculture system over here. It has been done in Australia and in the tropics, with a small farm built around a fish and water fowl pond surrounded by an irrigation network stacked with a whole variety of crops and fruit trees. Possibly it could become a demonstration model for the breadth of variety a smallholding could offer and the benefit of that type of farming for the individuals and the local ecology.'

I asked how Matt would afford the transition to aquaculture if it meant that he would have to cut down the time he spent on the money-earning elements of the farm to accomplish his aims. So many other farmers I had heard about were trapped into mainstream land exploitation because they were so in debt. He replied:

'Permaculture involves a small expenditure at the beginning, but then it's a very low-cost, low-management system. Over the first five year period, it's probably only a fraction more expensive than conventional farming.

'We'll introduce the design little by little. My girlfriend is going to run sheep on the farm for the time being. That will be quite commercial. Having said that, the farm doesn't compete with a pension fund growing barley fifteen years in a row on 2,000 acres. But that's a system which is unsustainable.'

After Matt had displayed so much energy in his work over the past few years it was hard to see him satisfied by a retreat to a hill farm. How would he cope with leading a life away from campaigns and city excitements?

'I'll still be a director of Worldly Goods and I'll do paid work

16

there once a fortnight. I'm also involved with Eco Environmental Education Trust, which is a whole different ball game.

'It arose out of a conversation with Pat Fleming. We wanted to set up a computer-based information system with a question-and-answer format.

'If you wanted to know about water, say, then the computer would ask you what you wanted to specify: water quality, river water or what ever. You could then, perhaps, find out what was in your tap water, what different scientists had said about it, maybe your MP's address to write a letter of concern.

'The aim was to produce an empowerment tool to say: "Look, this is what is going on in your area. This is what you can do." A shopper might see one of these machines in a building society's window and be pulled in.

'I'm also working on a low energy centre idea for an urban environment. I'm hoping to do that in Birmingham as a city-based scheme.'

Matt had been influenced by Bob Brown at the Wilderness Society in Australia who said: 'It does not matter what you do, you can still be Green. The real challenge will come when people stick to their profession but adhere to a very strict set of values about what they are prepared to do.'

'Do you think the system could change?' I asked.

'Yes, yes I do. We're not talking about fundamental change. For instance, look at my work in Worldly Goods. I'm a middleman, I'm in a commercial business, I'm wheeling and dealing. I feel very unhappy with some of that, but I'm doing it for the groups to extend their influence and for my living.'

'So capitalism isn't inherently bad if it's kept to a human scale?' I said, expanding the issues with a huge generalisation.

'Yes, I could buy that,' said Matt.

'I am also optimistic about the idea of digging your own patch. If it came to the crunch and the whole world crashed, people would be doing it within a month. They have got the ability to change that fast. They would collect their rainwater because it would not come out of the taps. They'd eat the tops off nettles because there are recipes. There would be a lot of suffering, but they could do it.

'The idea that we *could* change has a terrible ring to it, because it is as if people are aware that their lifestyle is wrong. And yet they still *don't actually* make the change. They know they are living out of the next generation's pockets or those of the communities in the Third World by exploiting resources in an unsustainable way, but they can't bring themselves to make the change. That to me is really worrying.'

Some people had adapted and they had inspired Matt.

'The strength of the Green movement is that it is seeding itself all over the place,' he said. There were self-help groups in Chile, there was the £1 million share issue at the Centre for Alternative Technology in Wales. In Australia, many Australians took the simple and pragmatic step of collecting the rainwater that piped down their gutters to drink. Apparently it was a simple exercise which required a water butt and some limestone to balance the acidity, not massive investment and a college degree.

The suggestion seemed absurd. I imagined that if the water from the tap was polluted, there would be little difference in the clouds. But his knowledge and experience struck me as the sort of thing I could pass on in my travels and writing. I could communicate these tiny insights to spread the word.

In London you drink water that has on average been through other people six times, and had some fifty-six chemicals added to it, Matt told me. There is no comparison with the water from the roof. Rainwater is distilled and purified. It leaves behind the oestrogen, the minerals, the chemicals which build up in our drinking water. So Matt could bottle rainwater on his farm and sell it as purer than Perrier?

'Absolutely. There was a guy who went to see Bill Mollison at the Permaculture Association in Australia after he finished college in Tasmania. He asked him for a good Green idea which could make him some money at the same time.

'Bill told him to lease the roof space of the football stadium in Hobart: "Go make them an offer for the down-pipes."

'So this guy went along and asked if he could put in a hundred year lease for the rainwater. The stadium owners looked at him in amazement because they paid to have the rain taken away through the sewage system. They leased the pipes very cheaply and the man bottled the water. Now he's a millionaire from selling it.

'Bottled water is worth more than petrol per litre! And yet every time it rains on London about a billion pounds of drinkable water goes straight into the sewers!'

What was the secret of this guru-like figure, Bill Mollison? Was he so much further along the path of discovery? Matt said he was a charismatic figure. He tended to get a good hearing just from the strength of his delivery. But he had also done his research and spent a considerable time to tune a coherent philosophy. 'Perma-culture is very advanced. It has been rigorously approached. The

Manual is a classic book. Even David Bellamy has said this is the best earth repair manual we have and David Bellamy knows his stuff, too.'

It was time for the train. We went downstairs again, over beautiful Turkish rugs, past colourful oil paintings. Matt could undoubtedly afford a rich lifestyle. In words that are true of so many committed, but human, converts to an ecological conscience, he sighed: 'But the real battle is on a very, very local level. It is whether I look out of the window in the morning and see the rain and decide I'll drive to work or not.'

My appearance transformed by a suit, I dropped Matt and his girlfriend off at Bristol Parkway station and headed for the Avon Environmental Centre with Andy. Andy picked up his messages. A councillor at odds with political opponents over recycling plans wanted help with his side of the debate. News came through of a Kent-based recycling scheme which had just folded because of the paper glut.

We took a tour of the centre, which Andy ran with another paid worker Mike Birkin, administration staff and an army of volunteers. We passed offices for a political hostage release charity and Matt Dunwell's Worldly Goods. Martin Fodor's playthings warehouse was cram-packed with goodies to stick on collages or bash into shape.

Outside, Andy had a brief chat with a woman who worked for Avon FoE Recycling Collective, FoE's recycling agency in Bristol. She was busy sorting high-grade, high-price paper from the dross in an enormous lock-up skip. It had been collected on rounds made by the co-operative to offices and homes nearby, on vans or with the help of their horse-drawn wagon. (Andy had worked for Avon FoE Recycling Collective before UK 2000 came up with the funds for his job. He worked 'as a glorified rag and bone man' for two years and was blowed if he knew why they ever kept on using horses— emotional attachment or something.)

Andy had a plan in the offing which he wanted to groom for Resourcesaver. His ambition within his FoE contract was to get a nationwide campaign up and running. He was going for office waste recycling and had consulted sponsorship experts who reckoned a multinational company would come up with a few hundred thousand pounds if they could see the principle in practice.

We ate lunch in a good old vegetarian cafe at the centre. It was the usual diner decked out in sprightly paintwork and offering a choice of herb tea with portions of tasty food. Andy left me with Pat Fleming, who had teamed up with Matt Dunwell on the Eco Environmental Education Trust.

Ten years previously, Pat had taught mentally ill patients how to count their small change in Woolworths and the like. She had a profound sense of what madness meant to society from this intensive care of six or so individuals, but felt compelled to look at the macro level. Her priorities changed.

'A healthy society should be able to contain its diversity and look after it. A lot of the problems we have are about relegating old people or those who have got Down's Syndrome or been made redundant.

'But what's the point of caring? What context do these sick people have if there is a breakdown in the ecosystems that support society? None of this matters if you can't breathe the air or be sure there will be earth beneath your feet in thirty years.'

Jonathon Schnell's book *Fate of the Earth* shook Pat with the need for radical change. Her life shattered. She left her job to come to terms with her own sense of doom and hopelessness.

'But,' I said, 'for most people the biggest step is not the conceptual one, but the practical one of leaving your job. How did you do that?'

'It was just an inner conviction,' she said, 'not a sense of what I should do. It was just in terms of my own integrity or beliefs. There was no career process.'

Pat signed on to the dole and took a week's course, called Despair and Empowerment, to learn to face the depression. She needed time to clear her own mind and pick up the positives rather than the negatives. So she went to the Insight Meditation Centre in Wiltshire, one of a whole network of sanctuaries across Britain, where she meditated for two months. Her retreat carried on when she moved to a cottage on Dartmoor.

There were two types of sanctuary, she said. One had a social and environmental conscience and the other did not. She preferred the former's 'engaged spirituality' because she felt she needed to have her spirituality applied to the outside world. The latter's search for some kind of nirvana was less essential. 'If it's not a spirituality in application, what difference does it make?'

Matt Dunwell had told me about the Barn, in Totnes, a Buddhist-inspired sanctuary where he had stayed. He was woken at five o'clock every morning to meditate, then bells were rung every fifteen minutes to keep him awake. He only got some peace when everyone went for an hour's communal prayer at half past six. Then he spent the day digging in the fields. Pat laughed. The picture Matt had given me had not been relayed to her like that.

I thought back to beer, curry and World Cup suppers, which

was what most people were tuned into. Life was hard enough without enforced labour. But Pat knew from her own experience the reason why people sought this stoic life.

'I'm less strong on the regime side of things,' she said. 'That has been imported from monastic life where people go to crack through into the deeper meanings of life by withdrawing from that life.'

I wondered how Pat might best describe the clear philosophical shape she saw to her life. I could sense the strength of her convictions and the depth of her experiences, but needed her to communicate them to me, to break through into simple language, that more might understand.

'I had to find what mattered in how I lived my life and how I felt about myself, how I was in the world and how I was with other people. I'm always cautious about taking on anyone else's rules or creed, even though it may be totally wise and practical. I have to check it out for myself.

'The tools from Buddhism and other spiritual practices were a way to quieten down the mental practices of the world. I realised I didn't have to be a victim to the realm of thoughts in which ideas are endlessly chewed over. That was part of being human in the same way as having arms and legs. I discovered the mind could be a tremendous resource and capacity but in some cases it was out of control and people went nuts or couldn't just switch off, relax and be with nature.'

Meditation improved Pat's concentration and focus on the potential for action and she shared her experience by leading workshops for others who had been through the same traumas. She felt she could understand what made people strong enough to act or else so weary they needed retreat. A sense of community was paramount. So many tentative hopes never came to anything because they were made in isolation.

This led her to a new role. She spent a year with the Eco Environmental Education Trust researching ways to improve public information about environmental issues. She concentrated on the best strategies for getting people to listen to the perceived problems, then to do something. In short, she was trying to evangelise. Matt's grandiose computer venture had rather been put on hold but the brief remained. Pat would research the best ways to improve public access to environmental information and use all her knowledge and experience on the psychological side to give people the strength to act on their convictions, to say: 'I will change.' Access and empowerment were the buzz words.

Most of Pat's time, however, was spent putting people in touch with each other, as a 'networker' on the side of the environmentalists and not the public. 'I'd be paid for it in America,' she said. She had talked to so many people about their work she got a fair idea of the scene, who was doing what, so she could put compatible researchers in touch with each other.

'I work under a sense that we haven't got time to waste. I want to prevent further mistakes by getting people to address the priorities. It's really a very critical decade in terms of setting up infrastructures that are really clearing things up. One of the most important precursors to that is dealing with the complexity of the information—about Gaian systems, ozone depletion, algal blooms. To many, many people it means nothing at all.'

But it all still came back to community, to dialogue, to seeds in people's minds, to communication. 'If I feel alone and cut off, I don't get excited or resourceful.'

Pat was a fan of Totnes, Devon, where she saw how like-minded people had managed to organise themselves and generated lots of activity which served the networks. Totnesians clung to the 1970s belief that the more people made their own shoes or grew their own carrots and traded, the lower they kept prices. And the community stayed strong.

Specifically there was the Totnes Green Pound Bank. Each member who transacted through it had a chequebook to pay for goods and services. They, in turn, could charge for their labours in Green Pounds. So the currency stayed in the community's circulation and did not pay for national scale economics with trucks to transport goods across the country or enormous advertising budgets masterminded from London. Any links, tenuous or undoubted, were severed.

Pat had become a regular visitor to Totnes and was contemplating a complete move to the area. But what of herself. Did she practise what she preached?.

On a practical, 'how-Green-am-I?' level, she had abandoned the car. Yes, she admitted she needed to use one now and again but she borrowed one from somebody else. 'If we shared cars we would reduce the number manufactured, a process with its own environmental impact.' Also, Pat no longer used black plastic bin bags—and the binmen tipped her rubbish into their truck all the same. Thirdly, she had an organic garden.

While a student at Edinburgh University Pat had helped run a garden share scheme for FoE. Grannies, with land they could not cultivate, were matched with students in flats who wanted the opportunity. They paid their rent in cabbages.

Edinburgh FoE shared their offices with SCRAM, the Scottish Campaign to Resist the Atomic Menace, and Pat went on a few marches to make up the numbers. 'I did not know it, but the seeds of activism had been sown...'

Andy had prepared for a meeting after lunch with Clive Attenborough of Community Service Volunteers, which was also in on UK 2000's arcane 'governmental non-governmental organisation'—a gongo not a quango. Clive had put feelers out for a CSV scheme to boost their public profile. Whereas FoE had established a public face through its campaigns, CSV was not allied to anything specific. They were looking to form an association with the National Federation of City Farms, based in Bristol, to give them a campaigning base.

Andy was in the midst of wooing the federation to take office space in his property, and also knew some of the people on the ground at Windmill Hill and St Werburghs City Farms.

We spent the afternoon at the farms, as visitors rather than on parade for the officials, which was usually how they did business. Both Andy and Clive would normally duplicate assessment work for the UK 2000 bureaucracy and had done too many such tours. They preferred a casual saunter through vegetable patches and billy goats with a quiet gossip over a cup of tea afterwards. We ended up in the pub.

Andy originally came from Hull, where he had had lucky schooldays. His comprehensive school did not officially stream pupils, but took some aside quietly and tutored them. He learnt to play bridge, through lessons given by the metalwork teacher, and enough of everything else to get him to university to read maths and physics. Unfortunately bridge was the start of his undoing. After fumbling with maths, he was thrown out of college twice.

He kicked around home until 1982, when he worked a harvest season in the local pea factory on twelve hour night shifts, so that he could leave England to travel with £1,000 in his back pocket. The winter, spent in Crete, was a magical experience. He lived rough in the hills with three other traveller types and 'the odd person we met in a bar and asked to come up to our cave'. He recalled that time sentimentally:

'The floor is very hard, but you sleep very well once you get used to it.

'I remember we had Christmas there. It was the first snow they'd had for ten years. We tried to cook a chicken but it was one of the village chickens and just too tough so we had a veg curry for Christmas dinner. Then a French-Canadian girl sang for us from the back of the cave.'

Andy had not made an overt commitment to Green living then, but he would sit on the hillside under a tree and think about his future.

'I had this fear, which returns to me, and it's an irrational one, a fear of being bored. So I took books because I was frightened that at some point I might not be able to cope with not having any information to take in at all. It's a habit you get into, rather stupid really, because once you sit on a mountainside your thoughts just fill your head and you find you're meditating.'

He worked in a cucumber factory that exported to Germany; some of his fellow workers were German hippies, told to pack only the ones mother would buy. The ones mother wouldn't buy were sold to the locals.

'A lot of them were blue with copper compounds from the fertilisers. The farmers in the market gardens would use powerful chemicals to grow peppers, cucumbers and aubergines. On the side of the drums it said in clear English and maybe one or two languages, but not Greek: "Use only with proper applicator." But they couldn't read that, they'd just stab it with a screw driver and pour it on. You think: "That's the influence of the First World on this, hardly Third World, community."'

At the same time there was a race of Cretan mountain men who would occasionally stop the village to drive their goat herds through. They didn't look anything like contemporary Cretans. They were tall and had light-coloured hair, blue eyes and a very athletic physique. They kept a few shaggy goats and sheep and lived a very simple, but very healthy, life. To Andy, they were impressive.

'They wore black hats with coins sewn on to them, had big moustaches and knee-length leather boots. They were very noble-looking and they were obviously living in harmony with the mountains. They were nothing to do with the fringe coastline which was essentially run by colonists like the British on tourism and market gardening—another example of the damage done just to that part of Europe.'

But Andy realised he was no great traveller and came back to England after that winter. It was 1983 and he went to Bristol where he knew friends who campaigned for FoE and collected the dole.

In an early Avon FoE project he organised a pilot car share scheme which cut twenty cars from the local commuter run. It concluded after six weeks that people liked to share a journey, but many of the shares were at the expense of bus rides, instead of extra cars. So the investigation stayed on the file.

Much of the work I saw on my visit to Bristol was still at a similar pilot stage. Andy defended this, optimistic that many initiatives would in future be adopted as examples of good practice, or at least be valuable personal experiences. As something of an expert at the age of thirty-three, his advice was:

'Doing it right in human terms is important. I am a little tired of the "what's wrong with the world" thing. We've got a pretty good idea what the problem is. It's mostly in people's heads if we're truthful and it's mostly to do with things like individual empowerment, where everyone gets the chance to do their own experiments and can value their own thoughts, their own personal growth.

'I say to people: "Look where the old school at FoE got us. They got us here. Look at the achievements they actually pulled out of a bag. For Christ's sake, look at them. They may have been just a bunch of hippies to you, but you are inheriting some very, very clever work that they did. You may be shrewd business people with a whole range of academic disciplines but you haven't pulled this together."'

Andy and I left to return to Montpelier. I chatted about his lifestyle in the back dining room. The sun streamed in, over the terraced houses behind and the creepers that lined the small back yard. To me, this life was laid back and it almost had glamour. Andy was his own boss, he worked hard but when he wanted, and he lived amid the rich trappings of a yuppie home of stripped wood floorboards covered in rugs, cosy rooms lined with bookshelves, 'real art' on the walls, answering machine. The loo even had a wooden seat, which is not altogether unusual, except that it was about the fifth I'd seen by the fifth day of my travels. Perhaps organic food does something to you.

Andy said the house was typical of FoE members he had visited across the country. His was decorated by his flatmate and landlady Dot, who collected trophies from her travels as a development worker across the globe. She was a thirty-something graduate with a keen concern for the environment and a research job at Bristol University. Dot's were the neatly ordered books in my room, classified in sections for travel, ecology and fiction.

When I put Andy on the spot about being a yuppie, he made four evasions before he finally gave an admission. Yes, this could be seen as a trendy home.

'But Jonathon Porritt's leaving party was about the most glamorous thing in years. Not because of the people but because it was on a river boat lent by Richard Branson.

'I'm quite alarmed when I hear some magazine say: "Is the Green trend over and are we all bored with Ecover?" But I expect it, after all they live on fads and fantasies and fashions.

'Within the broad Green movement there are lots of fashions developing all the time. There are new ways of looking at things because ideas are still young and history is being made very fast.'

I had a chance to catch a few potted definitions from Andy. I started with substainability and empowerment. Could he explain them?

'There's the old Indian definition, where in your dream time you ask the seventh generation ahead whether what you are planning is a hot idea or not, because it's a legacy for them to acquire. Nuclear power, for instance, would fall at the first hurdle.

'Another definition is to ask: "If I do this today and I do it tomorrow and every day for ever more, could I continue without harming anybody, the planet or anything else. And would it stop others from doing it as well?"

'And as for empowerment, when people are empowered, and it's usually for very righteous reasons, the word community comes in. It goes back to the simplistic roots of socialism: being in control of your own life. If you are empowered you feel more able, and less inclined to influence other people. You feel sufficient with your vegetable patch or community, you don't interfere with what anybody else is doing.

'If people already felt empowered by negotiation or by right then that would stabilise progress and we wouldn't have the horrendous armed conflicts we have.

'There's also the idea of the food chain. People want to participate but over the last couple of centuries this participation has faded away. They had food brought up from the pantry, the market turned into a supermarket and everything now comes in packets from God only knows where.

'The Green consumer movement has shown how to empower the consumer with the ability to ask some very hard questions about where that food was grown and what's been added to it, and to expect some answers. Otherwise the empowered individual could get back to the food chain by growing vegetables in the garden.'

As a parting thought, Andy confided his masterplan for sustainable empowerment in his own life. (It came from the same streak of northern thrift which motivated his Renault recycling scheme—buy cars near scrapheap stage and put them back on the road because 'the planet, having paid the price of a new car, doesn't get full value otherwise'.)

His father had recently died. He requested a quiet funeral without the family present, to which Andy agreed. But he still had to pay funeral costs of £500.

Enter papier maché coffins. Andy's plan was to construct papier maché board from recycled paper and turn it into coffins. He would bury the dead in an orchard and plant trees over them. Their death would be more fruitful than an end in a sombre cemetery.

CHAPTER THREE

Nigel Wild
& Red Herring Café

I could cheat no longer with the car. It was parked and I was home in Newcastle. I cycled over the other side of town to Fenham across the Town Moor. It was spitting rain but this time I was togged up to meet Nigel Wild in his allotment. He was at Newcastle University just before Matt Dunwell, on the same agriculture course though neither knew the other. Nigel went on to be president of the Students' Union on some kind of independent ticket and left with a strong urge to do something with his hands. He got his degree without ever having to grow anything.

I knew Nigel because he ran the Red Herring wholefood café and shop in Studley Terrace, Fenham, baking all the bread on a home-built oven in the back of his terraced house around the corner. We'd met at the Red Herring co-op on several occasions and at the 1989 Schumacher Lectures in Bristol, where Nigel proudly clutched a new copy of Masanobu Fukuoka's *One Straw Revolution*. I had always been humbled by his charm. He was one of those people you can't imagine would talk to you and, when they do, you are disarmed by the full attention they give.

The allotment was a new venture. Nigel took it on because of his strong desire to control every aspect of his life for himself. He wanted to learn to grow his own, and he did. He had not found full enthusiastic support from the other eight in his co-op, so he went ahead alone. In two years he had increased the area as he found he could cope with more beds, and as he climbed the rigid pecking order of old Geordie allotment hands.

He trundled along wheeling a barrow of eggshells and onion peel from the bakery. I tried my hand at the tricky operation of sprinting up a plank with it to reach the top of the compost heap. I did not have the knack or the muscle.

We chatted as the breeze flexed the branches of the trees and spattered raindrops onto us. I had no aversion to getting my hands dirty after the Glastonbury experience. So we weeded the runner beans together, propping them up on fallen twigs from the trees above (you don't get a shadeless plot first time round in Fenham—or a greenhouse for petunias).

There was a bottle-shaped clay mound behind which, Nigel explained, was a Scotch oven he built to practise for an outdoor craft event held in Stockport. He had been asked to build an oven and bake bread in it, all in one day. That was a dummy run. He stacked up bricks and coated them in clay, then stoked a fire inside until the bricks were hot enough. He raked out the burnt wood, loaded the dough inside and sealed the gaps. It had worked very well the third time.

We talked about the early days of the Red Herring when it was an anarchic experience set up as a café in an unassuming terraced house: a red herring because nobody thought it would come to anything. You paid if you felt like it, it opened when somebody felt like it. But Nigel developed some clear ideas about how to temper people's different perceptions of work with the need for a viable business to discipline itself. He also realised that he did not need to have very much to survive financially and materially in life. What was more important was an honest living, he said time and again, one which did not exploit other people. And he sang the cryptic words of Bob Dylan: 'To live outside the law, you've got to be honest.'

Finally, drenched, we picked brilliant blue borage flowers from a strong, healthy plant. Clean from the prickly stalk with a pinch, they would add colour to the night's salad. We loaded up the empty wheelbarrow with a box of fresh lettuces and rolled home for a cup of tea. The rest of the allotment holders were packed and off to their pigeon lofts.

Along the back lane, nothing could tell of the activity in Nigel's house except for a couple of Morris Minor vans, one dilapidated, the other spruced up in the Red Herring livery of deep blue and red. The yard was well ordered with salvaged scrap carefully laid into piles and on shelves. Nigel looked into the oven fire and poked about a bit with an expert eye before he shoved on a scuttle of coke. Under the house stood the red and green trolley used to wheel bread to the shop. It was beautifully restored with iron-rimmed wagon wheels.

Inside the back kitchen, three young men with long hair and white aprons beavered away at making bread. Dough for the next morning was under way, pizza topping on the boil, enormous pans in the sink. The roasted smell of fresh baked croissants emerged with a tray hussled out to stand.

Nigel grabbed two *pain chocolat*, filled a pot of tea and led me through the house. He had lived there over five years as a tenant—had had his thirtieth birthday there—but the landlord had never really tended his property. Nigel had taken it on since one of the landlord's friends had brought back their rent cheques saying he did not know where the man was any longer. Rent scrupulously went into a bank, but all repairs were well covered.

Pride of place in the sitting room was Nigel's printing press, surrounded, on the stripped wood floor, by typecases and sofas. One day Nigel would print and bind his own book of poetry.

In the back bedroom, there was evidence of his poetic ambitions and his motivation to take control. His home-made bunk stood six feet over the door the rest of the furniture was scavenged and rehabilitated from skips. The bookshelves spewed cookbooks and literature—William Blake, Tony Harrison and Herman Hesse—on to the shambolic desk.

After graduation Nigel had stood for union president as 'a fairly wishy-washy liberal' against the broad left. He was elected because the Agric pack thought he was a Conservative. But his sympathies developed a more anarchy-inspired approach, flowing from a tradition of independence, of not yielding to what you individually felt was wrong and of following principles not leaders. Nigel also began to make links between politics, philosophy and religion, which led to a fusion of Christ and Buddha, for their challenge to foolishness, hierarchy, bureaucracy and mis-representation, as well as in terms of the way they attained enlightenment. For a time he was involved as a poet with Red Umbrella, a radical collective for perfomers, then decided to try his hand at some of the things he believed in his heart, and opened a café.

'A political notion in your head,' he observed, 'is fairly meaningless unless you start involving yourself, interrelating and working with people. Only then can you translate it into anything concrete. For ultimately it is our behaviour by which we are judged.'

The early Red Herring, which he had mentioned in the allotment, served simple, cheap food and suggested a price. The till consisted of a bowl marked 'Contributions to World Revolution'. The thought of it and the crunch in the brown flour *pain chocolat* launched Nigel into an exposition of his inner convictions:

'Realising the limitations of what I have done has not necessarily prevented me from doing things, because part of the problem is that you are always trying things huge and vast and to change the world, and you actually can't do anything because you are not looking at where you are and how to feed yourself. I think it is really important to learn to do these small things that aren't significant beyond your own life, but through which you learn about how to do real things.

'I suppose I wanted to learn how to feed myself in an actual and a more philosophical sense. It seemed important to start to take control of myself and of the environment and of the resources you need to feed yourself. In a way, I developed this notion that food is important because it is the point of contact between rural production and urban consumption; and food, if there is nothing else, is the one thing common to all because we all need it to eat. We also need resources and land to be able to produce it; and land is fundamental to political change. Land ownership, land reform, understanding the political implications of ownership, is really important.

'Food is a way into that. Being able to feed yourself is really important if you want to change things. If you can't feed yourself then you are always dependent on people who feed you. It seemed to me that if we fed ourselves then we would be in a stronger position to say: "Look, this is wrong, we know it is wrong, we demand change and you cannot starve us out because we can actually feed ourselves."'

Behind the rhetoric and the principles, however, there was very hard labour. Nigel had to get up in time to buy from a fruit market, then returned with the sacks on his own back to stock the kitchens. On one side there were the hippies who said: 'Oh well, if it opens, it opens, man.' On the other were the organisationalists who wanted everything done by rota. For Nigel, stuck in the middle, there seemed to be no happy medium which reduced his personal workload. However he had shown himself he could start a business, had confidence he would never need a conventional job and, after nine months, he left the first café collective to its fate.

An interlude abroad beckoned. Nigel had read *When I Walked Out One Midsummer Morning* by Laurie Lee, *Homage to Catalonia* by

George Orwell and *For Whom the Bell Tolls* by Ernest Hemingway.
'Let's go and meet some good anarchists in Barcelona...' He set off
with £40 in his pocket and got as far as Paris, cold and penniless.

'I stayed for a bit and thought: "Barcelona is an awful long way.
I don't know the language. Do I stand much chance of earning
any money? Maybe I ought to stay in this cheap hotel I know." The
hotel was in the Latin Quarter and used to be a brothel; a
remarkable and peculiar place where some Brazilian transvestites
lived. I got some work there and on some film sets.'

After a time Nigel came back with his tail between his legs. To
recover, he went on a cycle holiday in Cumbria and came across
the Village Bakery in Melmerby. He heard there was a job going as
a gardener at their smallholding, was taken on and went home to
sort himself out. He bought a split-screen VW van down a back
lane in Crook the day before he was due in Melmerby and drove
over with all his possessions. For a few months, he slaved by the
bakery oven because they needed him there more than outside,
but he would often drive over to Newcastle to support anti-
Falklands War events.

After the summer, Nigel cut loose again because he wanted to do
his own thing. The Melmerby set-up was not completely his style.
Andrew Whitley and his wife Liz had moved from the south where
Andrew had worked on the BBC Russian Service.

'They are committed to growing things organically and they also
want to earn a fair bit of money and do well out of it. They are
trying to create some sort of alternative like quite a lot of people of
their age or their background, in that they managed to opt out
because they had quite a lot of family money behind them that
enabled them to go live in the country and do "alternative" things.
That's a sort of criticism but I defend them when people criticise
them by saying: "Look if you had some money, what would you
do? Is it better to have a job in a city and earn lots of money and
drive a big car, or is it better to try to do something you want to
do? Set something up in the country and grow organic things?"

'We are all faced with that decision. You can simply abdicate by
not making it and say: "I do what I do and that's all there is to it."
Fair enough, but if we have a choice, if we have some privilege, we
are bound to exercise it. If you perceive yourself to be privileged,
and I don't think you perceive yourself to be so if you aren't, you
should accept it and try to do something beyond that, to put
something back into society. After all, it is society that conferred
privilege on us originally.'

As a start, Nigel had his van to do little removal jobs, and his connections with peace and political movements enabled him to continue catering at marches and meetings. He then moved into the large house he has occupied since.

'I think we were all on the dole at the time, a bunch of mates living a life of some lack of seriousness. But I got delivered a sack of flour occasionally from the Watermill, near Melmerby, and started making some bread for the household. It got to the point where I was making it every day. I was getting up quite early, about seven o'clock, to make a batch of dough, then meditating for half an hour or so, by which time it had risen and I put it in the oven. I made about a gas oven full every day and started to sell it to people round and about who heard I was at it. They would knock on the door and buy a loaf every day.'

I asked how far Nigel thought there was a 'bread for empowerment' movement because making your own bread had become quite a fashion statement in itself: 'And I do make my own bread, man.'

'If you are making choices about the way you want to live, you obviously want to eat good bread. "Am I going to be influenced by the cheapest, crappiest manipulations of our society and eat manipulated, adulterated bread or am I going to go out of my way and find organic, stoneground wheat and make some bread myself?"
'That is a step of liberating yourself from the shackles of consumer society and it is also a therapeutic thing to do. You get these very few simple ingredients and that process of kneading is therapeutic. You are developing the gluten in the bread in a transformation.
'Those transformative experiences are really important. Creative. One of the "healthy" things about it is that it is almost immediately consumed. It's not like creating something which then exists as a work of art. It is simply a part of the process of living and you are engaging yourself as part of that process. You are not producing to admire. It's getting into your own life.
'For a long time I have struggled to do that. The van was part of it, learning how to make it work. I was terrified—I am about practical things. If I don't know how they work, I am daunted by them. Making the van work was an important thing for me. Learning how to make bread was too. Getting up to make it every day was an apprenticeship. I practised the simple skills of kneading and moulding with two hands and making batches. It gave me that discipline I think you need if you're on the dole—because you are able to get into a lifestyle of some debauchery—

the discipline to give yourself direction.'

While Nigel made his daily bread, he thought about ovens, like the Scotch oven at Melmerby. If he had a bigger oven, his catering business would take off. So he just went ahead and did it, built a coke-fired oven with a flue plumbed into the original house chimney. He was not even sure it would work, so he didn't go to the expense of engineering central heating out of it as well.

'I always wanted to heat hot water for the house. When I was building it, friends who were helping said it would be quite easy to heat water. I said: "Exactly what do I do, what do I need and how much is it going to cost to do it?" They couldn't answer those questions precisely and I was really skint. I suspected it would cost in the region of £150-200 to do it properly. Maybe on another occasion when we have to do some repair work, or maybe if it becomes something that I decide is urgent, it would be sensible. It would save some money, some energy.

'My ambition would be to have a small steam turbine as well, to generate electricity for the house but, you know, one day...'

The ideas did not stop there. Nigel had seemingly followed through every aspect of his life. Our discussion ranged over the global economy and how Traidcraft tea was still a cash crop which kept the poor developing world yoked to our consumption, no matter the responsible wages. Land ownership was the original sin because it began with 'the expropriation by violence for profit as the only end'. The tax system was iniquitous because you could not opt out of paying for defence, so Nigel would go to court refusing to pay.

I sat intently and listened to the blond, bearded Sheffield man, dressed in fashionable, Maoist blue overalls.

'I think food was a good starting place because it's so important to all of us. One, because unless you're well fed and healthy, there's very little you can do anyway. A lot of people are unhealthy because of what they eat, and we are manipulated by the food industry. We have this great underclass of people who are ultimately at the whim of producers and kept in unhealthy conditions by the food that they eat, which is over-processed, over-fatty, over-sugared and has all sorts of detrimental effects on the heart, teeth, liver...'

Was that a symptom or a product of the fact that people like to eat sweet things, I asked, considering which came first, the supply or the demand. How had people let themselves be manipulated if there was not a tiny grain of pleasure in sugar coating? Were they just lazy and too weak-willed to avoid it?

'I think it's an addiction to some extent. Where does that start? Well, sugar, say, is a very fast hit. It's something you can taste and then it has an effect on the bloodstream and then your mind and body. The effect is significant and profound. But, in a way, it's a part of the culture of immediacy which we are locked into, where we have come to expect things at once. The types of food which are adulterated with sugar or whatever—to give them that little lift, that "I want it, I like it"—are a reflection of the society we are. In small quantities there's maybe naught wrong with it, but that immediacy is sapping.

'With regard to cultural forms, like pop music, it's very immediately available and appeals to the lowest common denominators. Instead, I think it's important to be involved in things. Culturally, the pop music scene is dreadful because people, for the most part, seem to be observers of it. We are encouraged to live vicariously through other people's experiences, through the experience of pop stars, even through the experience of soap operas. It denies the values of our lives and elevates the value of other people's.

'And that's what the royal family is and to some extent Parliament, and the whole governmental system. People perceive that Westminster is the centre of power and that we, outside it, are powerless. We perceive things change there and this doesn't give us the confidence to change things ourselves or to feel that we have a voice in our own lives and in our own communities.

'To some extent sugar is part of this. It's something we are addicted to and it does not do us any good. That's both a symbolic thing and actually damages our bodies. Also, it has a whole iniquitous system and history if you look at the agriculture on which it depends, which is thousands of miles away in the Caribbean and came from the slave trade. It's a cash crop that demands vast monoculture systems which employ slave labour, still very poor labour. Where the cane sugar has been superseded by beet sugar, you have vast estates that, although they don't grow sugar anymore, won't let the people have access to the land. It's a whole pernicious system, an example of the global market.'

Nigel put two and two together like this until he had quite a large score. But he turned it into a positive career move by supplying local people with the most locally-produced foods. 'It seems to me that if we can live from the things that grow around us and near us, then that's sensible,' he reasoned. 'If we are in any sense adapted to the environment in which we live, then probably we ought to be able to live from the things which grow around us.' There would still be the distinct possibility that thousands of miles from his bakery children would be starved because their families were denied land to grow crops, but his living would not support that system.

'"Think global, act local" is a really significant maxim. It's so easy to think the dreadful things that happen thousands of miles away don't involve you and you can't do anything about them. But there are things you can do, which is one of the reasons I never went abroad to try to help poor people in a conspicuous manner. I felt: "Well, I can't go to Africa and pretend to help people to eat food. They have lived in their country all their lives, they know more about it." But maybe what we can do is change people's attitudes to food and to life and to other cultures and to their land and the whole number of interrelated attitudes and patterns of production and consumption. They are all related.

'It's quite a struggle to get people to understand that the notion of land ownership is, in actual fact, not a God-given right. It's one that greed has managed to foist on the majority, by the minority, through violence. The interesting thing is, if you start to challenge it even in peace, like the Diggers, then generally the forces of the state, which has its interests in ownership, will come down upon you.

'Significantly, in the 1980s miners' strike, people started to say: "Hold on, this government is starting to close down our pits that are our living and if we allow them to do that we will just be on the scrap heap. Is this a sensible way to do things?" They started to exert their influence and say: "We have some say, we need some say, this will affect the lives of thousands and thousands of people directly, never mind indirectly." Then you saw the gloves coming off and the state being what it will be. On behalf of the landowners, honest decent folk in the police force ultimately had to "kick the bastards in". It's very brutal and very crude and doesn't do anyone any good.

'People think they can insulate themselves from it but it isn't doing anyone any good. Certainly, it isn't doing our children any good. If only people could start to think not for their children, but for their children's, children's children, like the North American Indians' notion of affecting the seventh generation hence. If you can't think in terms of how long it takes to grow a tree, how can you possibly cut a tree down?'

The solution to immediacy? Count the costs in the first place. Nigel was the only person to appeal with a non-statutory objection to the latest plans by the Tyneside Urban Development Corporation. He went along to say he didn't think their plan for quayside marinas and yuppie flats was in the best interests of the people of Newcastle. He stood up to be counted, saying he was appalled by the way the corporation had usurped the role of the local planning authority.

Nigel had a vision to match that of Matt Dunwell and I longed to bring the two together to work on their urban low-energy centres. Nigel was trying to obtain council consent to develop a forty-acre site up the Tyne into a farm and alternative business centre, christened the Tyneside Environmental Centre. Matt was looking to put his energy into something similar. I asked Nigel the same question I had asked Matt about the difference between thriving on cosmic visions and the humdrum world of daily baking. Did it matter?

'Accepting the repetition of it is important. Lots of us want, and are educated to expect, powerful and exciting positions. We are expected to have some degree of control and be important and drive fast cars with car radios. There is an assumption that those things are important but, in the end, I don't think they are. They are just a sop to our ego. The notion of importance is nonsense. All we are is someone who lives in a community who has a little bit of say.'

There was no hint of bitterness. Nigel was not important, he said. He did not have status, no. He had a role to play in the community which was quite 'exciting': he was active in the local anti-poll tax campaign and had been invited by his vicar to join the parish community forum to discuss local issues.

'What would you do if you had a million pounds?' I asked.

'Probably sit down and think about it. Perhaps buy land.'

The first stage towards a million was laid in the Tyneside Environmental Centre. The idea was to withold the proportion of tax spent on defence and plough it into a city-based alternative centre: allotments, self-build homes, homoeopathy, bakery, energy efficiency. Backing came from the council, UK 2000, and lay foundations. But it lay in the future and, he said, 'The further we go, the more we are dominated by men in grey suits with grey hair.'

My time had come again to put something back. In Bristol I had intended to spend a day out on the horsedrawn paper collection, but I had overslept. I was determined not to miss my morning in the bakery.

It was all go by six o'clock. The oven was stoked and the dough set churning in massive mixers. Nigel and fellow co-op worker

Matthew Davison, on baking duty for the morning, were joined by Andy McDermott, an artist at the polytechnic on summer vacation. In three weeks he had picked up the basic kneading and shaping skills and become very good at weighing out the ingredients for pizza topping.

This was my task for the day. Peel the onions, peppers, mushrooms, boil in tomato sauce, add industrial quantities of 'a pinch of pepper, a sprinkling of oregano'. Every now and then I made the tea and washed the pots. I suffered badly from curiosity and had to sample everything—even the concentrated butter scooped from the EC surplus mountain and rolled into the brioche dough. There were sweet loaves with nuts and currants, sour loaves made with yeasts which naturally occur on the flour, some risen in a traditional scrubbed wood trough which Nigel had acquired.

I greased tins for loaves large and small. I learnt to roll a ball of dough on each palm into rounds, all smooth with the raggedy bits collected underneath. These were then dunked in a topping of poppy seeds and squelched into the tins. A trayful was whisked off on experienced shoulders to the rising room beside the oven.

The croissants had to be crescented with a splayed hand and dabbed with glaze. The buns had to be counted for the regular sandwich orders and for pot luck on the day's sale. Breakfast, a ritual of cheese and onion in bread rolls, was eaten on the run. I sat down to guzzle bread peanut-buttered from a Meridian tub so big any nut fan's eyes would light up. Andy reluctantly gossiped about his art. His current fixation was air vent covers. He himself had breathing problems. The covers became the statement, no matter where they were fixed. Some had even been fitted with sound to billow out. This was good stuff for the imagination as we shared our time together.

Time flashed past. I had an appointment down the Westgate Road in Elswick to meet Richard Adams, head of New Consumer magazine. I said my goodbyes and bundled slices of pizza into my panniers for distribution at my next port of call. Nigel warned me against 'bonk'—when your blood sugar drops, you become exhausted and just want to die on the spot. The cure is apparently Mars bars.

CHAPTER FOUR

Richard Adams
& New Consumer magazine

Richard Adams set up New Consumer after rigorous thought and consultation and had run the operation for eighteen months. The idea was a logical follow-on from the work he had pioneered in ten years as head of Traidcraft, a Third World trade and aid business importing commodities on non-exploitative terms, a market solution to the imbalance of trade with the developing world. Richard's immaculate credentials, which included an MBA, and the first book published on alternative trade, were sufficient to attract massive grants from the likes of the Joseph Rowntree Charitable Trust when he came up with a new think-tank idea to influence capitalism for the good.

New Consumer was deep in research to rate British business on its ethical awareness and social responsibility. Some four researchers sought to document precisely the activities of the hundred biggest consumer-oriented corporations in Britain and plot the values that they lived by. Did Marks and Spencer donate money to political parties? Did Whitbread have women on the board? Was Nestlé linked to Third World exploitation? And gambling, alcohol, animal welfare, arms trade, worker democracy...?

They wanted these answers brought out into the open, not so that they could knock companies with values they abhorred, but to give the consumers power to make up their mind. The results would come out in their own magazine and through two books, edited by John Button (see Chapter Nineteen). A new agenda would be set, beyond the mere biodegradablity and energy-efficiency of carrier bags and hair driers. By differentiating policy, they would force businesses into a comparison of their values. A simple extrapolation of ideas would look to social justice on a global level. Yes, the washing machine only ate ten kilowatts of electricity every year, but was it built in a sweatshop in Cairo by penniless orphans and mothers? Were the profits spent on developing high-velocity rifles?

Richard was modest and unassuming for a man with £30 million of Third World imports through his books over fifteen years at Traidcraft as well as distribution contracts for all Greenpeace mail order merchandise. He had a beard, a beat-up Volvo and an office full of young workers who despaired at his dress sense on official occasions.

'I never regarded my life as being particularly Green,' he told me as we sat in the busy office full of computers and potted banana plants.

'In terms of personal motivation, it's more a question of social justice than environmental concern. As is being shown increasingly, the two are inextricably linked because of the nature of consumerism and exploitation of resources. Essentially, I regard myself as a socially-concerned person rather than a Green person, although Greens would argue you probably can't make that dichotomy.

'I do not label New Consumer as deep Green, either. We rejected the idea of using Green in the title because we wanted to ensure that the public understood we were talking about corporate responsiblity in its broadest sense and not just environmentally. If you look.at the world you can't but notice that probably the major force in it is materialism and the way people's materialism is serviced through business structures of one form or another. The last couple of years in particular have shown how, more and more, the market economy is gaining strength and squashing the opposition. In many cases it is relatively value-neutral. So it's really not a matter of being deep Green or holistic but of saying: 'How can one redirect people's perspective if their desire for material goods, affluence and so on is being catered for by this extremely powerful, very sophisticated system?'

Richard calls himself a social entrepreneur as opposed to a conventional one: the social entrepreneur champions social concerns by doing things that need to be done and has a very definite view of how they should be done.

Richard's life's work became the redirection of society, particularly where no direction was apparent, and he aimed to achieve this by charting the current movements on his sophisticated scale. Money itself might be neutral but how people spent it asserted their values. If they chose the cheapest they might have taken the selfish line—the environment might suffer or workers be exploited to cut the price.

In Richard's case, his Christian convictions swayed him towards these questions. But business had lessons which he could not avoid learning.

'I am not a business man, good grief no, although I would not reject some of the systems and disciplines that business has evolved to regulate, moderate and produce goods and services efficiently. But I think these have to be tempered with a perspective of human potential, human development and what is appropriate for society both globally and nationally.'

It all began while Richard was studying at Durham University. He showed his acumen and mischievous nose for enterprise by hiring out dinner jackets to social aspirants. Prices were cheaper a train ride away in Newcastle, so it was worthwhile paying an impoverished student to spend the afternoon up the line fetching the order. Was this the training for an ideas man? He subsequently went on to run a greengrocer's shop with a partner who had contacts across India and Africa. They imported vegetables.

'You get more credibility if you put your ideas into practice. It was fairly easy to work out in the abstract that small farmers in the Third World couldn't get a decent price for cash crops; but that there were quite stable and much higher prices in the First World. So the answer was to try and get their produce over and sell it here.'

The difference was in the producers they dealt with. In the first year they imported 200 tonnes from small farmers. At the same time they sought to inform the customers that their trade supported the small grower. But peppers and French beans and mangos were perishable produce depending on very strict control by agents and shippers. There proved to be little potential for a 'message'. Ethnic craft was much easier to bring over, had immediate cultural references and did not go off in transit. After two years, in 1975, the vegetables were superseded by craft products. The company, called Tearcraft, sold these products through wholesalers and commission agents and gradually developed mail order catalogues. Initially Tear Fund used resources in the Third World to ensure a market for the community-based businesses it sponsored.

'You could highlight how much more effective it was using money in that way because it essentially became seed capital. This would generate continuous employment instead of just one-off donations that fed somebody for, perhaps, a day, a week, a year, and no more.

'The other aspect of it was that, once we got on to the establishment of Traidcraft in 1979, we were also trying to demonstrate alternative kinds of structure and business practice, participation and so on.'

Pioneering business organisation became a most important aspect of Richard's work. He wanted to explore how things could be done ethically, and make his business so successful that conventional, reactionary wisdom had to take note of a viable alternative. Interestingly, he spurned co-operative ideals and the thirty-odd models which co-operatives can create.

'I have never been enamoured of the co-operative concept. It always had problems raising capital and I really don't think anybody's got to grips with the difficulty about liability. Also, I have doubts about who is ultimately responsible for getting things done, given the range of skills and types of abilities you need for a co-operative. It can be a group, it can be delegated. My own feeling is that job flexibility needs to be matched with specific job descriptions. You can operate like that in a co-operative but it is very difficult.

'I think they are good for some things but they start to creak, especially in worker co-ops which get beyond the coffee-break circle, when you have to differentiate jobs.'

I wondered whether he worried that co-operative working tended to operate in low-paid businesses, where workers had to exploit themselves to survive. Where did money and reward come into the equation?

'Oh, I think the two are quite separate. You can have co-operatives where the members pay themselves extremely high salaries, because they are successful, and ones where because the business is not successful, or because they choose to, they live on a pittance—just as in a conventional company.

'People have to be motivated to work in a co-op. I don't think negative motivation, like rejecting consumerism and materialism, is a good basis on which to do any type of creative, constructive venture. I think it has got to be much more positive than that.'

These strong views corresponded with my own feelings. They restored some dignity to the dirty words of enterprise and initiative. Richard recalled the experiences of his own start-ups, none of which were co-operatives, to explain.

'I have never had a job where I have earned more than fractionally above the average industrial wage, although I have been fortunate in that, since 1973, I have been running companies that I, in essence, started. I think it's a matter of approach. Even if you have got a successful company, I think you can still say there is a limit to the amount of money it is reasonable to make. You don't have to be motivated by money to run successful organisations. To me, differentials are quite important for different jobs but in some respects, the more important in conventional terms a job is, the more intrinsic, intangible rewards it has. If you're the managing director of a large, multinational company, yes, you have responsibility but also you have quite a lot of rewards in terms of privileges. You get the ability to control and direct and influence and make your mark. These things nourish you.

'I think that's one of the problems of co-operatives. They don't encourage people who really have a feel for what I would call 'social entrepreneurship'. Co-ops are good for some but a lot of people go into them because they think they will enjoy the working situation, the association with colleagues, the ambience of the thing. In some respects this becomes more important than the results.

'That's great. If you like that, it's probably the job for you, but it's very difficult to tie that up with certain types of social objective. Sometimes there are things you have just got to get on and do. People in co-ops must put up with an immense amount of pressure. It's not necessarily the hard work, but my feeling is that in some co-ops people draw the line at that. They say: "I didn't actually come for this. To me, personal development and space are important and I am not prepared to sacrifice that." Given the imperfect world that we live in, I see no option but to sacrifice a whole range of things that one would ideally like to have at the moment.'

Traidcraft mushroomed at a time when value-led business was in its infancy. Before, it was charities that sold with a message. Americans sold refugees' crafts after the war. Oxfam started selling goods from Hong Kong projects after 1962. So Richard represented

Tearcraft and Traidcraft at all the conferences in the field from 1975 onwards.

'Very exciting, particularly if you see a convergence of views towards the position you think is the right one,' he said with an embarrassed splutter of laughter. 'It might not have been so exciting if it had gone the other way.'

The debate was linked to two views of the global economy. Richard held that if you could not sell it, there was no point in making it. You had to advise the manufacturers on what products and designs would appeal to Western tastes and support a transition. For instance, in Bangladesh the jute market collapsed as synthetics took over. You had to moderate the effects of such a change but recognise, at the same time, that traditional jute could never be used in modern machinery because you could not guarantee its regular tensile strength.

The opposing position was that you should educate Westerners to appreciate the wares of another culture so that it could survive as it had done in the past.

Traidcraft achieved more than modest success and Richard's role became more that of a 'safe pair of hands' than that of an innovator. It was not his style so he left to write a book on his experience, with no other concrete plans. Along came the offers. Economist Paul Ekins, the first Green Party secretary, wrote to him in the course of a project on how one expresses alternative consumer values in contemporary society. He knew of Richard through his administration of the Right Livelihood Award, the 'alternative Nobel prize', which had at one time considered Traidcraft for an award. Meetings followed and Richard took on half a week's work besides the book.

'Where did you get sufficient strength of character to say: "I'll just chuck it all in and something will come up"?' I asked.

'It doesn't work like that,' Richard replied. 'It's not a sudden decision. You don't suddenly decide not to be phenomenally materialistic. It just sort of accumulates. I didn't have any worries that I would be able to do something myself. I did actually draft something to stop me and my wife worrying but I never actually referred to it because New Consumer came along.'

With typical gusto, Richard threw himself into defining what exactly New Consumer stood for and the best strategy for getting results. His views on other established pressure groups like Friends of the Earth and Greenpeace gave me an insight into how to place them in a political context, and how he saw New Consumer.

'I think there is a bit of a problem in that Greenpeace have cornered the high profile, public action, campaigning area quite well, and Friends of the Earth would find it difficult to compete. But on the other hand, from what I can pick up from the people who have left FoE recently, they seem to be losing the experienced environmentalists who could have given it a much more in-depth, researched, informed approach. If you were trying to draw up an overall strategy plan for the environmental movement you'd say: 'Here's Greenpeace—high-profile, direct action. They'll hit ICI, who may have done some good things, but forget about them and hit the nasty things.'

'Down at the other end you've got to have a body that's respected by people in industry who are trying to make changes but are finding it very difficult because of the various pressures. They need support for what they are doing. Now, at the moment, I don't think Friends of the Earth sees itself in that role. It still sees itself as primarily campaigning with public action and so on.

'Neither of them seems to be very good at giving information to the public. We get tons of calls to New Consumer on environmental issues. Some of them are referred over from FoE or Greenpeace. We're not an environmental organisation at all, but they don't have the resources to answer public queries on the environment.'

Richard's strategy for New Consumer dictates that its research base is so sensitive to the facts and responsibilities of mega-business that no one can criticise its research. New Consumer has to be authoritative.

'Certain patterns are emerging. You take two companies, both started within a few months of each other in 1970, both roughly the same size: the Virgin Group and Iceland Frozen Foods. Both around £700 million turnover. They started in different sectors: in the case of Virgin with Richard Branson as an individual; with Iceland, two entrepreneurs who set up a shop in Oswestry. Iceland have grown steadily and organically, staying in the same market for the last twenty years. They have policies—very good policies—worked out on the environment and personnel issues, on participation by staff, industrial democracy and so on.

'The Virgin Group is virtually nowhere on all of those issues. It doesn't even have written policies on most of those things. It's grown an entirely different way: by acquisition, buying different bits and pieces here and there; obviously struggling to integrate a diverse range of stuff; still very dependent on its chairman and erratic in its policy. That's something that's come across. There are companies, some asset-stripping and acquisitive, which grow by buying other companies, in a whole range of diverse operations. They tend to come out really rather poorly in terms of their social responsibility.'

Richard was a strategy player. He had thought through the points of contact which he would have with the outside world and directed himself against those he felt needed to change. He was, like all those I met, a man of vision. And it did not stop there. Richard had cannily installed another of his hobby horses in the same building. Shared Interest was a hybrid venture capital business, run by experienced executives from the sharp end of that industry, with the goal of supporting enterprise in the developing world. Lenders would give money in a similar way to their charity contributions, except that they could stipulate some return in the form of interest or a lump sum. A model for the future?

CHAPTER FIVE

Bruce Marshall,
organic farmer

The Russian poet Irina Ratushinskaya added a new meaning to the term 'concentration camp'. She wrote whole books of poetry in the Gulag but, denied paper, she had to save them in her head until her release, whereupon she scribbled them down and published them to acclaim. I was incapable of sustained thought as I began my cycling. But then I had the luxury of constant excitement, rather than cell walls.

I heard crickets sing on the verge. I counted five different red flowers, five blue, five yellow. Wild roses, honeysuckle and bramble flowers dusted the hedgerows. I smelled them coming as I passed. Foreign cars flashed by from Germany, Holland, Belgium, with all sorts of lives to imagine contained inside. The landscape expanded and I felt small and unbelonging.

I pedalled North, the sun on my back, uphill on the map. Thoughts filled my head, as Andy Moore had predicted, so I sat and unwound them on the beach at Lindisfarne, my first day's resting place. But more gushed in. At a speed of only twelve miles an hour, my life had taken on a different pace.

I watched the evening tide flood over the muddy harbour. I was trapped until the causeway cleared at ten the next morning. The sun glanced off Lindisfarne castle and I watched the day's last beam dance down the coast behind a train of clouds and away, off

Longstone lighthouse, off Inner Farne island, Bamborough Castle, Seahouses and a smoky coaster chuntering south.

The fishermen picked up their tenders, moored in the deep channel and skulled for home. One by one the catches were landed and they shoved their rowing boats from the pier, anchors balanced over their sides, attached to a long rope. Thirty yards off they tugged their line. The anchors bucked overboard followed by juddering chain and the boat was secured, afloat for the morning. I've seen the trick performed many a time to keep the rowing boats afloat off the beach and secured for the morning. But only here. A minute detail of tradition lives on.

I wondered about changes: in the passage of time, passed in the sweep of my pedal, in the drag of the wave as it combs the pebbles back to the sea. These islanders resisted the pull. Everybody here knew everybody. They knew them well before they could speak. They remembered them in their graves, all family names together. Here was strong community spirit.

I could have sat out all night scribbling such thoughts. It was comfortable—but my legs were so stiff I could not move.

Next morning was brisk, grey and Friday 13th July. The Farne Islands looked near, which is taken to mean bad weather is on its way in Lindisfarne. I squeezed everything back into my small panniers: army surplus sleeping bag. Camera. Tape recorder. Notes and addresses. Map. Two changes of clothes, one verging on smart. Trainers. My one prized possession was pair of sandals bought in Minorca. They were made from two thongs of leather stitched into the sole, which was originally a car tyre. All Minorcan fishermen wore them. I thought they were a canny use of modern technology and also elegant.

I left the cottage where I had spent the night and tightened a few nuts on my ten-speed racer, a British-made Viscount emblazoned with the name KP Crisps. With the weight strapped to the back, my tent on top and anorak within easy reach, I teetered off on my groaning legs, buying a stock of Mars bars to fend off 'bonk', and a copy of the *Guardian* at the Lindisfarne Post Office.

How far-sighted I had been at Glastonbury! I must have been standing on a ley-line. I got a puncture just before I reached the Pentland Hills at about five in the afternoon. I had misjudged my fitness, and the average speed I could make up and down the windy route. I changed the inner tube and headed on, swearing loudly at the lonely sheep in my frustration, to make the wide open spaces seem less desolate.

Fine irony. I took a sign marked 'La Mancha' and was tickled by thoughts of Don Quixote. Both of us tilted at windmills. I mean, cycling sixty-odd miles in a single day to visit a hill farm! Oh, the anguish of idealism turned to action. Cycling was not for softies; they passed at regular intervals in their otherwise empty cars. I felt alone and roughing it.

As I coasted through La Mancha hamlet, what did I find? Two pantechnicons laden with peat bound for the gardens of England. And the campaign to save peatbog habitat was in full swing. There were ogres in modern La Mancha, not windmills.

After ten hours in the saddle I arrived at Tocher Knowe, Bruce Marshall's home. I had dreamt all day of what I'd find and had made a mental note to compare the land on approach. His was an innovative organic farming technique, I had heard. Would I instantly tell where his land started? No, the grass was healthy and high, but not visibly different. Tocher Knowe, up the valley from West Linton, was a long, low, wooden house perched on the edge of a deep glen. It could have been on the Serengeti Plains of Tanzania but for the round heathered hills.

It was dusk. A Land Rover and two other all-terrain vehicles were parked in the drive and surrounded by deep grass. Bruce greeted me at the porch with a gravelly hello and a firm handshake. Then he turned to curse Tigger, his disobedient labrador, as it slid up the Persian rugs in the hall. Bruce was dressed like a country gent—tweed jacket, check shirt, beige trousers, suede slip-ons. His hewn face was an indeterminate age and he refused to put a figure on his years. I reckoned—old.

Two cans of McEwans on an empty stomach later, I pitched my tent in the long grass and died—or would have done but for the flies and a slalom of tussocks. Both dug into my flesh whichever way I turned. The 'Environment' *Guardian* became my extra protection and a ready blanket.

I rolled over at nine next morning face to face with Colin Thubron, the travel writer, in adamant discussion on the page opposite. 'I think the imagination has something to do with it,' he had written about sleeping alone. 'You feel you have a rich world to inhabit without the presence of other people.' I nodded. Silence is hard to break sanely when you're on your own, even with your Rosinante and Sancho Panza (or in my case, KP Crisps). 'I think people on their own eat less than others,' Colin reminded me, to my chagrin.

I was out of iron rations as a result of the consolation feast I had eaten when my tyre went flat. There were three fruit shortcakes left for breakfast. After polishing these off, I freshened up in the gushing burn and went to rendezvous with Bruce for a tour of the farm. This was back to nature. Wordsworth would have approved of the transition. A month from the claustrophobia of exams I had taken to the hills and ended up cleaning my teeth in a brook.

Bruce had taken over his 1,500-acre hill farm in 1971 after years as a tea-planter in Ceylon. He was packed off abroad after agricultural science at Oxford, too young to run the family estate. He returned with a British wife he had met in Kenya, built Tocher Knowe, and managed the family's 5,500-acre estate before he split it with an elder brother who took the grouse moor and the big house.

Bruce could have afforded to let his farm bumble along. He had 'other money'. The sheep virtually kept themselves and there was game for sport. The scenery was life-giving, the wildlife wild—an idyllic outdoors for his children, good subject matter for Rosamund's landscape watercolours.

Instead of the 'softly softly' approach, Bruce chose to improve the land. After all, he reasoned, that's what he would have been told to do on any other type of farm. He scattered minerals to correct soil deficiencies and satisfy grazing animals' needs. He worked hard to balance the acidity and alkalinity of the ground, and sowed clover.

Then he discovered earthworms. They thrived on the refreshed soil, chomped through it, making rich loam and further encouraging the clover. Different species munched the tussocks and the rushes. Some even feasted on the hundred-acre peat bog, which was a nuisance as regards the improvement plans. The worm corpses doubled the nitrogen content of the soil created by the clover. They were a godsend to fertility, a natural and sustainable source.

So Bruce took to 'harvesting' earthworms from behind the ploughs of bemused fellow farmers who did not seem to want them; then he sowed them in strategic lines across his land and left them to it. The buried treasure was marked out on a crinkled map spread out on the dining-room table before we set off. The worms burrowed and multiplied about 3 metres downhill each year, Bruce reckoned, and shaded the chart accordingly. Worms and clover added fertility so more sheep could graze. The sheep topped up the soil with manure.

'Everybody treated the symptom of poor drainage, whilst ignoring the disease—infertility,' Bruce said and rolled up the chart again. The pasture came to withstand drought better on his farm and there was less run-off after heavy rain. Bruce had collected together data to prove it and to defy the experts, the establishment, the scientists.

'After all,' he said, 'I am a trained scientist too, whereas most agriculturalists are just chemists. I am also a hill farmer and everybody else has just mined their hills until they are no good.'

He has written a pamphlet, which, in best eighteenth-century tradition, is called 'New Energy Creation and Conservation Policy to Benefit Farming, Woodland and Wildlife World-Wide and create Rural Employment using Nitrogen-Fixing Plants, Earthworms, Native Trees and Shrubs'. It spans his discoveries, his many experiments and his inspirations beyond the earthworm phase, in a rambling collection of jottings full of charm and eccentricity. They were arranged in no apparent chronological order, though carefully dated.

Bruce discovered that moles followed the worms and thrived. They helped to churn and drain the soil.

'Most people seem to consider molehills a sign of bad farming,' he said in a paper to the 1984 International Earthworm Conference in Cambridge. 'I don't think that. Other people worry that moles will kill off the earthworms. I don't know any predator other than man that has killed off its main feed quarry.'

Charles Darwin's little-known work, *The Formation of Vegetable Mould through the Action of Worms*, published 1881, is quoted:

'The plough is one of the most ancient and most valuable of man's inventions; but long before he existed the land was in fact regularly ploughed and still continues to be thus ploughed by earthworms. It may be doubted whether there are any other animals which have played so important a part in the history of the world, as have these lowly organised creatures.'

To this Bruce adds a postscript: 'The objectives of ploughing seem to me to be to bury the surface litter, and to prepare a perfect seed bed—all done by earthworms for free.'

The miscellany continues with a quaint letter from the Rev. Gilbert White, at Selbourne, dated 20th May 1777:

'Worms work most in the spring; are out every mild night of the winter, as any person may be convinced that will take the pains to examine his grass-plots with a candle; are hermaphrodites, and much addicted to venery, and consequently very prolific.'

Bruce makes the observation, in a paper of his dated April 1988: 'Colleges and quangos have backed a system of perpetual semi-starvation for hill sheep.' His figures show a doubling of livestock on his farm after the fertility was reinstated: more lambs born and better land available to feed them because of the worms and clover.

The thirty-six page study does not stop there. After his conversion to earthworms, thanks to his daughter's biology teacher, he started to restock trees native to Scotland. A forestry expert said it was not possible, as he stood with his back to healthy birch planted by Bruce's grandfather. Bruce planted alder, rowan, willow, oak and

51

wild cherry instead of the ubiquitous pines favoured by the experts.

'It seems to me hypocritical, to say the least, that environmentalists in this country are noisily urging the Third World to stop cutting down their rain forests. We cut down the Caledonian forest just to survive, may never know how many species were destroyed in the process, and were only saved from erosion and desertification by our climate. Yet even with our present knowledge, we are planting up vast areas of our hills with alien trees, rather than trying to re-establish our native species.'

Bruce's improved soil now bears thirty-six species of native trees, shrubs and creepers planted as shelter and windbreaks. They were not planted in regimented rows but sprinkled across the lie of the land. And finally he had come to a scheme marked 'still experimental': hedges against inflation. He was cultivating rows of willows edged with hawthorns, wild rose and gooseberry.

'I've planted ancient hedgerows,' he joked, 'like my "ancient natural woodland". They are so species-diverse that they resemble the natural state which evolves in these habitats after many, many years of evolution.'

The organic approach has left game to multiply, too. The woodlands, worms and clover have given excellent food and shelter for pheasant, partridge and black game. And in years to come the firewood from windfalls and coppicing alone could keep somebody in employment, Bruce claimed.

Bruce and I climbed into his brand new Land Rover to tour the farm. He brought along a spade, a yellow plastic watering-can with a litre of water, and several jars marked 'Poison'. We passed my tussocky camp by the glen.

'It's virtually a Site of Special Scientific Interest I've created, you know. The Nature Conservancy bods come and look at it but won't register it because I practically made it myself. I fenced it off from the sheep and cattle and all sorts of wild flowers self-seeded. Even trees like hazel are growing back, which shows nature reverting to ancient woodland.'

'A prophet is not without honour, save in his own country,' he added with his gravelly cackle, and we sped off. New Zealanders and Australians come to visit often. Bruce has been courted by the French media, but only slowly have the British become interested in his work. An agricultural college lecturer once visited, and returned with his students but was never seen again. The Scottish Woodgrowers' Association had called recently, and

Bruce had decided that the time had come to 'go public' with the results of his labours.

We were out on what once was a hundred-acre bog with peat ten feet deep. It looked like any other meadow, green with grass and clover, grazed by cattle and sheep with the odd purple Scotch thistle. I dug with the spade, and saw a cross section of dark topsoil about eight inches in depth. I'd even sliced a couple of worms. Below, peat lingered but, I was told, it receded an inch downwards every year.

Bruce filled the watering-can with water and added poison in droplets—formaldehyde. He prepared a small tray with fresh water, then sprinkled the potion on a square yard and we waited. One by one worms stung by the liquid wriggled to the surface and were rinsed off in the clean water. Bruce then popped them in a jar to take them to another experiment. He wanted brandling worms to eat his household waste in vats beside his home. He had shown them to be capable of working on shredded newspaper and was now on to whole sheets. Could a heap of worms pull apart whole *Daily Telegraphs*?

Bruce apologised for the fact that there were not more worms. In three minutes we had fifteen and a couple of excuses reminiscent of fishing trips: 'It must be the dry weather, they sink in after a few showers...'

We walked across the old bog. I could see that it had shrunk. Trenches, where once shepherds had dug winter fuel, were stranded in meadow pastures. Clover had spread across and sphagnum moss, the plant base to the peat, was sparse and pecked out by rooks. On one side Bruce pointed to some trench drains, which had been one conventional attempt to tackle the bog. But when it rained even a tenth of an inch, soil would pour out of the gully and into the stream. So Bruce bunged them with rushes.

Across the boundary fences I could compare different techniques. The golf club had gone for ten-foot trenches around a stunted spruce wood with cosmetic native trees at the fringes. The next farm had spent a fortune on pipe drainage, and had planted a field for straw. Above, dull brown grass bore witness to undergrazed, less fertile land only yards from Bruce Marshall's pasture.

In the gully on Bruce's side, a copse was planted with trees that made up ancient woodland. Alder, rowan and gean grew in scattered array and the only concession to modernity was the plastic tubes to protect early growth. It would be glorious to return in ten years to see the land speckled with woods.

Bruce's latest wood was planted on a more adventurous site, one which the experts considered completely and utterly impossible to root trees in. It was an alder wood that had been set into the bog edge at a cost of thousands of pounds.

'It would only take six years to establish my ideas on a working farm,' said Bruce. 'Other farmers could pick up the experiments I have done immediately, and no time would go to waste. Pioneering has meant slow progress, but the benefits are already being reaped. The farm now employs two men, a shepherd and a cattleman, as well as fencing and tree-planting contractors, which is more than it has ever done.'

We went back around the valley through the farmyard to look at the hedges. They were beautiful, with roses lining the fence. Bruce reckoned he could re-use this fence after ten years, when the willow and gooseberry would be enough of a prickly deterrent to keep the tups and the ewes separate. So he would halve the cost of new posts and wiring, in the long run.

A tree nursery had been established, too, on a sheltered corner. Cuttings for the fringe shrubs were struck straight into the soil. Alongside, Bruce had two whimsical experiments. If the experts insisted on planting alien pines, he would plant gum trees for koalas and bamboo for pandas. The plants were doing rather well. He had also planted an American pine with a high turpentine content to see if it would grow in Britain. The wood could be used for outdoor purposes for which we have hitherto had to import timber.

'As you can see, I'm crackers,' Bruce told me. I'd had a few hints from his pamphlet. Quoting Schopenhauer, he had written: 'Before it is accepted, each problem runs through three stages. In the first stage it seems ridiculous, in the second it is opposed and in the third stage it is self-evident.'

CHAPTER SIX

Laurieston Hall Community

S o to my next stop: Laurieston Hall, a community of twenty
adults and ten children. I heard of them through the book
Diggers and Dreamers, a directory of communal living. Few of the
fifty-seven communes on the list lived overtly by ecological
principles but Laurieston Hall was one of the longest established
and revolved around the self-sufficiency idea. Besides, communal
living and shared resources are intrinsically green; and, the
resulting lifestyle and culture explored much more radical
alternatives to relationships with self, soul and soil.

Diggers and Dreamers said that all heating was from woodstoves,
with home-made hydroelectric power, a loch and a wood-fired
sauna with pond plunge. These were the fruits of eighteen years in
the commune business. In 1972 three couples bought the hall and
twelve acres. The expanded group, less all but one of the originals,
now owned 140 acres and had access to Forestry Commission land
for firewood. Their mansion had become a summer centre for
alternative holidays and conferences, and could sleep seventy-five;
it had been christened the People's Centre by a more Marx-inspired
generation. Laurieston Hall averaged 750 visitors a year.

The communards had, in the main, retreated from draughty
dormitories to caravans in the woods and to converted outhouses,
and Laurieston Hall no longer called itself a commune. But this re-
establishment of personal space had led to disputes and a

breakdown in the total communal living ethic. So in 1987 the commune—which meant pooled income and meals together— changed its legal and management structures to become a housing co-op. Each 'member' became economically independent and income-earning work (at the People Centre which caters for visitors) was reorganised on an hourly paid basis.

Gilly (no one lived by a surname), forty-three, a Lauriestonian for seven years and previously at Crabapple commune, told me why she had backed the change, as we sat one late afternoon out on the lawn.

'I personally didn't want to spend all the time struggling with the basics. So much of a commune is agreeing which food to put in the fridge, which loo paper to buy, because there are so many people with different ideas. So many pass through and new people are at the basics stage again. I didn't want to talk about that sort of stuff for the rest of my life.'

Members I spoke to pointed to the length of time most people stayed there and the number of new recruits as a sign that the integration as a housing co-op had been successful. Their shift, though in defiance of the purists, somehow paid off. The Hall has a large, stable community and each adult earned about £2,000 a year. 'But,' said Phil, 'we live like kings.'

Everywhere I had visited I could draw on one or two key points to distinguish a different focus. Nigel Wild had given me arguments about land reform I would never have considered on my own because of his link through food to cultivation and on to the ownership of the soil for seed. Richard Adams had concentrated on business and the impact it could have. At Laurieston Hall the talk was all of energy.

Residents spoke in terms of the energy they had to channel into a certain project, say milking the cows or starting a pottery course. It was an essential part of the psychology of empowerment towards action and a way to express ideas of self-development, a focal issue in the communal life. Members might feel able to commit something to action at a certain stage in their thought process, motivated enough to carry their project through the community decision-making process and place the burden of responsibility for their idea on the group. New members were accepted somewhat in terms of how much energy they might bring. It was motivation, initiative channelled to the values the community held dear—co-operation, harmony, the personal journey of life and the need for improved relationships.

Insurance was the modern con-trick, the key shackle to an old money-based system. If the community had security and lived for the present nobody would need to plug into any financial circuit. If you died the community would support the family. (Already I was blurring the different meanings of 'community'. The notion of a commune, a village life, even 'community' of interest.) Financial insurance perpetuated a solely cash-based value system, whereas the Laurieston Hall folk realised that more was necessary to live a full life, if you were a widow or crippled by accident.

Much of their workload was turned against this circulation of money, into community service. The members' workshare scheme saved them money instead of earning it from others, because everyone turned their hand to building maintenance, hydroelectric power, cheese-making, gardening. These tasks either added value or saved cash expenditure.

I rode up to the house along a tree-lined drive, marshy meadowland on one side. A tractor criss-crossed the hay on the other. Pot-holes made for a bumpy ride but I was impressed by the stone sweep of the stable block as I passed, its ramshackle grandeur given a New Age twist by a hand-painted sign advertising pottery and bright gloss window frames. A courtyard acted as shelter for the massive two-acre vegetable garden beside the house. Raspberries and strawberries burgeoned in neat ranks. Peppers were grown in the poly-tunnels (signs in the porch suggested times to volunteer for picking duty, or good feast areas for excesses). Four beehive-shaped log stacks followed the grass verge.

The hall's impressive classical portico loomed up ahead. I had a twinge of excitement at the country house opulence, the estate-owner style of the good life. Ten cars, at least, surrounded it; a climbing frame and swings, paddling pool and giggling children covered the ragged front lawn.

I walked into the house and was immediately welcomed. Tea from the urn in the hallway was thrust into my hand, space found on a tumble-down sofa by the fire and a left-over supper placed in front of me. Phil, a summer school administrator, sat and chatted. I had arrived as a 'dance and live music' camp reached its climax with a ceilidh in the village. Phil and I shared the *Guardian* quick crossword as we drew breath from our day's activities.

People bustled everywhere. The dancers had got used to the hectic social round at the hall and expected that I was a long lost friend of the community, while the community members thought I was a visitor they had not yet met. I smiled, said hello and worked on the permutations: if two's company and three's a crowd, then what was this?

Two enormous wooden satyrs glared down from the fireplace. Helpers shoved trolleys of washed-up plates back to the dining room. Course leaders (notice no terminology hang-ups here) with clipboards pranced to their workshop sessions. A gaggle followed. Somebody stood halfway up the sweep of a huge staircase and rolled a cigarette.

'Get out, Charlie!' An official boomed. I practically dropped my baked apple in fright. A gang of seven-year-olds bustled away, led a merry dance by my namesake, who initiated diminutive visitors into neat ways of getting attention. Folk dancers with only a week's experience on his patch were a doddle.

I asked Phil a few of the questions everyone always asked. Where did he come from? What did he do there? Why did he move? How long ago? Was it different from 'normal' life?

Yes, community living was different from town or village life because you extended your family by about thirty people. But everyone had come from, and still belonged to, the pattern of a car, a house, 2.2 kids and a dog. They had not gone through some magic transformation to make them in any way special. There was tolerance and intolerance, there were conflicts of interest, there were certain ideals which some of the group shared.

I asked how the community could afford it? How did they ever come to consensus decisions at their weekly meetings? What were the current important matters under discussion? Phil rocked back and grinned through his beard. He was thirty-something, with wife and two children, aged seven and nine, and entirely content with life at Laurieston Hall. It showed in his healthy tan, his lean fitness and a little twinkle in his eye. But he would not be drawn into any philosophising. There was nothing really weird about a community. It was all straightforward. The money came from visitors. If you needed to earn it you worked at the People's Centre. If you did not, because you drew other income like a retirement pension or a job outside the community, so much the better for you.

Phil came from near Doncaster, became a town planner in Cumbria and chucked it all in after the 1976 heatwave. An office was a prison that summer. His work in Whitehaven, near Sellafield, had become a 'yes sir, no sir' process, rubber-stamping the nuclear authority's plans. The folk he had gone to help were being picked up and shaken, as was his idealism. Phil left. He was single at the time and could take each day as it came. In the summer of 1977, he set off from Cockermouth to walk around Britain. The weather was not a repeat of the year before. He got to the Cotswolds one evening and pitched his tent in somebody's garden by the stream. It rained incessantly overnight and in the morning he poked his head out for fresh air to find he was stranded on an island in the middle of a swollen river.

The adventure was rained off but Phil learnt a lot about other people from long evenings in the pub. One night a man came over to his table and asked, straight off: 'What do you think about the Fourth Dimension?' Phil was treated to a learned introduction to the subject before he could make up his mind.

Back in Cockermouth in the early 1980's he relearned how to sleep in a house, and odd-jobbed. He met Annie and ran away to Wales with her. They married and lived on a barge, tramping the canals. Phil wrote a book of towpath walks then enrolled at the School of Peace Studies at Bradford for an MA.

The way he told me, I could not tell why he had chosen the course. It was about nuclear disarmament, the Third World, social oppression and personal peace—not quite Phil's bag, I thought. He had been particularly drawn by anarchy and its philosophy for the individual, and he left halfway through his studies to try it out, as did others who were motivated by the ideas sufficiently to go put them to work. His aims were characteristically down to earth: a return to village life and a house he'd bought in West Cumberland when he was still on the council career track. But he soon found that the village he had left no longer existed. Neighbours did not sit out on summer evenings and chat across the high street. It was a main road now. Institutions like the pub had been updated and villagers were distant from one another. Phil and Annie, with two young children, felt a huge gap in their lives. They wanted people to communciate with, not to pass the time of day with. Annie came to a Laurieston Hall course and discovered a place where they could do that. They could love and hate all sorts of neighbours. They approached the Lifespan, Wheatstone and People in Common communes, to test their own commitment and the range of lifestyles. They settled for Laurieston Hall and arrived in 1987 for good. The commune was dissolving, the housing co-op structure was being put into place. 'There was lots of personal emotional stuff going on at the time,' Phil said. 'Some people had not really wanted the change and left. Annie, who had initially been so keen, found it very unsettling for about a year. So did the kids. But I don't think they would want to move now.'

The bones of idealistic love & peace communal living were laid rather bare by the rearrangement in the commune itself, and the heady talk had come down to very specific objections. 'Oh, lots of emotional blackmail goes on,' Phil admitted. 'And, yes, obviously you get to like some people more than others. Everyone weighs up how much power they ought to give away to others.' There was no disguising the fact that other people always ruined an individual's utopia, but probably ultimately for the better.

So how about the financial aspects? Each adult paid £59 a

month to Laurieston Hall housing co-op, which paid for housing, electricity and heating. In turn they earned £2.75 an hour working for the People's Centre. Phil was in the collective of five who ran it, others cooked or organised specific courses. Life got complicated when the communal aspect came into the equation. Each adult was supposed to spend 'half their time' on workshare, working on the co-op projects to keep costs down. After approval of new members, the most frequent arguments were over the meaning of 'half your time'.

As I talked to Phil I was introduced to about twenty people. Phil collected visitors' money for the week. They paid from £8.50 to £15 a day, depending on their earnings. Nobody spent anything else unless they went to the pub down the road, so it was a cheap holiday.

I finished my supper and took my bags in search of a bed. The so-called Warm Room, over the kitchen, had an empty double mattress, so I put down my gear. Margot, my new-found room mate, suffering with a cold, offered to show me round the house and grounds. 'You must see the loch,' she said and insisted on showing me. I was weary and doubtful, but, a stroll away through the hayfields stacked with bales by the afternoon dance class, the woods opened on to a lake speckled with waterlilies and rushes. A thousand bullrush spears caught the evening light. In the clear light, stark granite was picked out on green hills far away. I, too, spent the rest of my stay telling newcomers: 'You must go and see the loch.'

Margot showed me the hydro-electricity plant on the way back. It hummed smugly. The initial £6,500 investment had paid itself back double already and it would last many years still, with hardly any maintenance. Two saunas were pointed out to me but I never got as far as Peter's straw house in the trees, hand-built and reputed to have its own central heating.

We went back for evening activity: campfire songs, stories and dancing. Over a cup of herb tea Alice introduced herself. She had arranged my stay, as she was in charge of the week. She was also the only Laurieston founder still living at the hall. She had brought up her children there. Alice told me how she had battled to educate them herself. She wanted to be responsible for what they learnt herself, instead of leaving it to some local authority school. But inspectors pried, and threatened legal action. Alice might have set a precedent and they could not allow that, so they forced the children to join the village system.

Under the stars, accompanied by the crackle of the fire Kim, an old pirate of a course leader, led the tunes on his mandolin. He sat draped across a chair, strumming and sucking a roll-up cigarette. Evi, a native of Fife, doctored the fire. All couples hugged each other. We chanted 'We are at one with the infinite sun' and a song in the round with everyone taking turns to celebrate an absent friend:

'I love a man from Birmingham deep down in my heart.
She loves a man from Birmingham deep down in her heart.
Deep, deep, I say, deep deep down,
I say, deep down in her heart...'

It was all too much for Callum, a lanky, self-conscious adolescent down from Orkney with his mum. 'I'm neutral,' he said point blank, when it came to his turn. There was no such full stop for the happy campers. They sang on: 'He-ee is neuter-al deep down...'

Inhibitions were obviously left behind in county of origin, because out came poetry. The atmosphere was unusually right for reading out loud. The group was moved to silence by T.S.Eliot's 'Four Quartets'.

All I could think of to follow this was Stevie Smith's 'Croft', a poem of nine words, and a few Japanese haikus. But I did not dare to extemporise.

Next morning, I stirred early as room-mates padded about—and also because there were no curtains. In fact, I never saw any curtains. Perhaps someone wanted to get back to nature's rhythm with light at dawn and dark after sunset. I lay, a contented cat, as sun streamed through enormous high sash windows, and browsed the Laurieston Hall yearbook, a self-conscious booklet printed at Lifespan commune, near Barnsley. It introduced me to community life and reviewed the last season in school magazine style:

The Secret Diary of Adrian Laurieston, Aged 41³/₄

June 25th. Talk about work with the co-op. Discussion brings up issues about relative wealth, support for those less able to work, expectations of communal work in relation to money-earning work, difficulty of finding interesting part-time work locally. Look for dynamic action plan to achieve social utopia. Decide to circulate a questionnaire.

Feb 26th. Drains blocked. Wish I were 11 and knew nothing about drains.

April 24th. Linda asks the co-op to sell her some land. Wonderful discussion (though not for Linda) renews spirit of co-operation. Ponder mysteries of consensus decision-making.

John, a new member, broached the subject of the psychological side of communal life:

'We don't really have a consensus of will for a forum to work out the more personal issues, or so it seems to me. But that may change, like so much has here over the years.
'I hope so because I like things to be out in the open and am interested in clear contracts that people can make concerning their feelings and dealings with each other.'

David explained his own angle:

'Something of the art of living here is to do with balancing my desire to try different jobs with my guilt about never being effective enough.
'My past year has been about coming to terms with living more on my own. Funny coming to a community to feel safe enough to do that.'

The articles were soporific stuff. Then, on page 8, I found a decidedly more saucy discussion, which took in some of the fringe benefits of communal living.
'Fears of losing my long-term relationship with my lover seem to be unwarranted,' it said. 'We've now been involved in one triangle of lovers for over a year, and a separate triangle for eight months.'
I racked my brains to remember my old geometry lessons and avidly read on.

'I'd like to be able to hand on hints as to why things are working better now. My long-time lover and I have been close to being involved in other triangles several times in the nine years we've been together, and dipped our toes in the water too, without much success.
'There's an element of keeping my fingers crossed—touching wood—and a feeling that if I stop doing this the fragility of the triangles will collapse like a pack of cards...
'I'd be happy to talk to anyone on a one-to-one basis who's also in a triangle (or two). I'd prefer to do this than make general statements.'

My interest aroused, then dashed on the verge of detail, I pondered why the intimate discussion of such many-sided relationships were to be conducted on a one-to-one basis.

Kim, the dancing pirate mandolin player, had been up all night.

His consultation papers and suggestions were laid out at breakfast for the visitors to choose a morning's worth. The invitation to learn more Macedonian folk dances or frolic in the sun was peppered with sweet nothings about peace and love: 'You are all such wonderful people I love every moment spent in your company. We've come a long way together this week...'

I joined the game-players in the garden and warmed up with one called 'fear of farting', invented in the small hours after smoke-gets-in-your-eyes time round the fire. Each player in a circle took turns to crawl between the others' legs. They then massaged you as you passed.

Team sport was in store, with competition, winners and losers, and Kim ordering us around like schoolchildren—even though he had a 'Greens are Gathering' sticker on his guitar case. Two teams lined up in combat and marched to play a unison 'paper, scissors, stone' game after a huddled discussion of the team's chosen symbol. The grand finale round had no such pow-wow, to test the team's group consciousness. One, two, three... our team almost unanimously chose stone. But then again a cheating nod and a wink is easier than telepathy.

The rest of the morning was spent in rehearsal for the farewell ceremony. We learnt a Celtic ballad: 'May the road rise with you, and the wind be always at your back... Until we meet again, may God hold you in the palm of her hand.' Everyone took parts or followed on a musical instrument. Talent was not a prerequisite. What mattered was that everyone enjoyed their attempt at playing together. Everyone congratulated everyone else, rather in the style of a love-in. This was the culture of involvement, of which Nigel Wild would have approved.

The weather was fine, the pace easy. Lunch was leisured in the open with pots of vegetarian bake, fresh lettuce leaves and currant salads. We drank more tea. I sat with Sheila and Louise, both single parents, and Gilly. Bronwen and the crèche arrived for hugs and lunch with the mums. Bronwen was a sixth-former from Wales who was child-minding as a holiday job. As a paid worker, she was half on the inside of the gossip. She was allowed through the doors marked 'Residents only'. Bronwen had decided that Laurieston Hall was just like *Dallas* the soap opera. She had identified the local JR and Sue Ellen while chatting with Gilly, but Gilly had never seen *Dallas*, so could not go further. Probably just as well really.

Gilly at least clarified the issue of the love triangles over our giggly meal. 'A triangle is only true if every one of the participants relates sexually with every other,' she said.

Phil joined us but would not be drawn on love matters. I asked him and Gilly how they saw their future. Could you get old at a commune? No problem: a fifty-year-old tax inspector had just joined up. He could not do back-breaking work but he had plenty of life for the living and his savings went a long way. Lively debate followed. 'You have to start from where you are and live for the best option at present,' said Phil. 'I don't see any point in slaving away for forty years to find myself too old to enjoy time with my family and too decrepit to get out and about. The commune would support my wife and children if I died or whatever—or, at least, it would give them a chance to support themselves.

Sheila agreed. One parent families were trapped by government benefits and the wilderness of city life. You had to spend so much time looking after the kids, you could not start to scheme a way out to work and pay minders. At least with the support of such a community, mothers could help each other out with child-care tasks and share the community work. There would be more sympathy and more time to learn or use your energy to enrich your life. 'It's not the physical side of having something to show for it, it's not having a huge house, mowed lawns and two cars,' she said. 'I want to have a rich life with opportunities and time to do what I want.'

Phil's children had undoubtedly benefited. They had constant playmates and a cosmopolitan flow of people who took an interest in them. Louise painted a picture of old folk renting their homes, constantly worried whether their finances would last them out. At a commune, the shared wealth would do that and little excess would be needed to generate further cash. 'And there is loneliness, even in a rest home,' I said. I had stopped off at my grandmother's on my first day's cycle. She called it a granny dustbin.

'In a community, age is irrelevant,' said Phil. 'Each person has a chance to help out, whether it's baby-sitting or imparting some of their experience. In "retirement" you are marginalised. No money from pension funds or insurance can compensate for this. And the idea of scraping about so that you can leave your kids something is a ludicrous product of the last twenty years.'

Yes, they lived like kings and queens at Laurieston Hall. But how were they treated outside? It had taken sixteen years for them to organise the first public dance at the village hall. Was the community a bunch of incomers? A few years earlier, at the peak of the AIDS scare, they had organised a week for gay men to come and holiday in the peace and isolation, behind high hedges. An outcry ensued in the national and local press; 'Lock up your sons'. There were letters of support from the neighbourhood and much 'leave them to themselves' sentiment.

The sun shone all day and the community busied itself for the evening ceilidh: band practice for the Lauriestonians, while visitors picked meadowsweet, bracken and buttercups to deck the village hall with swathes of greenery. A thrill went round. We were a family before a wedding, class-mates before sports day, dressed up and on our best behaviour.

I dashed to the loch for a plunge before supper. Bliss, I was alone. Carried away by the inspiration of my hosts, I dived into the clear water in the altogether—a sexual experience! Flooded with freshness, I floated back to the dining room in my trunks. Its doors and windows were wide open and a scented breeze blew in.

My room mate John and I walked down the Hall drive to the village hall together. He'd grafted as a social worker since unhappy days at Cambridge and, with the mortgage paid off, he'd decided to come clean with himself and packed in the job. He intended to lead a new life of diversity. He played the flute and the piano. He sang. He ran the local FoE group; and courses like those at Laurieston were cheaper when you were unwaged.

The village had never seen its hall dolled up in such a way. The landlord at the Laurie Arms opposite could scarcely believe the crowds and he was positively beaming. The party was joined by over fifty local friends and neighbours, as well as Dutch and French tourists who happened upon the gathering. I had a few whirls with Marie-Claire and improved my French with her ten-year-old brother, talking about the World Cup and skiing. Monsieur invited me for a beer after the dance and I explained, in rusty terms, the ins and outs of the community just a couple of miles from their lodgings. Madame, an immaculately dressed petite woman who had patiently smiled as I wrenched her arm on an earlier dos-à-dos, was aghast. 'Mais il n'y avait pas de vêtements de Woodstock,' she reasoned—but there was no one dressed in Woodstock fashion. Her children had seen some, but had thought nothing of it. Outrageous dressing was no longer subversive to them.

Monsieur and I supped our *demis* and discussed with typical French flourish (I had imbibed sufficient to pass as a Frenchman by this stage) how *communautes* in France had *champignionné après les années soixante*—mushroomed after the 1960s. He did not subscribe to *l'idéalismé*, but he could recognise holiday yarns he could dine out on for weeks.

The family offered to give me a lift home, with one aim in mind: a good look at where I was staying. We drove along the windy back drive, which I explained had been planned to reflect status. The Hall came into view. *Alors, c'est un chateau*, cried Monsieur. Madame gasped.

The dance troupe was leaving and the next week's guests were yet to arrive. I offered my services to the Laurieston Hall team who were due to spruce up the place, and was volunteered for a multitude of chores. I could pick flowers for arrangements in every room, I could clean up the dormitories. Ali took me off to the woodshed with a chainsaw, leather chaps and visor, and I promised to help Toni with the milking later in the afternoon. The log supply was down but there was plenty of seasoned wood to chop. It had been stockpiled from earlier expeditions to thin dead trees from Forestry Commission land. I stacked, Ali sawed, until we had a good heap. She was so fresh-faced I took her to be just older than me, but she was my namesake Charlie's mother. He had lived with his father in East Anglia for some time. Now it was her turn to take charge of him.

To keep out of the way, I cycled to lunch at the Laurie Arms down the potholed drive. The pedals on KP Crisps flew after my three-day break from the saddle and I sped with the delight of knowing I only had a couple of miles to go there and back. Bliss. It was so hot and still and dry, I could almost hear cicadas sing.

Amazingly, the barman was none the worse for wear after the night before. He busily sold sweeties, cold drinks and ice lollies to village children and to the road gang at work outside. I had come to say goodbye to the friends I had made on the circle dance course. The Skilling family sat patiently, with Sheila and Louise and their girls and boys, waiting for beefburgers and pop. Ron Skilling was an environmental planning officer for Knowsley Council, Merseyside, and a veteran of more than one Laurieston holiday. He and his wife Helen were also practitioners of Transcendental Meditation at Skelmersdale—Helen was convinced that its benign power pulled the crime rate down. I had not had a chance to introduce myself to them properly while we played 'fear of farting' the previous morning, and they had not realised I was as much a voyeur of communal life as them. I had again stumbled on people with a completely refreshing view on the Green movement and a positive livelihood to match.

Ron was responsible for an impressive scheme in his Merseyside borough to cultivate vast areas of council and private property which had hitherto been industrial waste land. The borough had a large stock of industrial land awaiting development. But the recession in Liverpool meant Knowsley had a good many years to go before demand for land outstripped supply. Ron had put together a plan to plant out spare plots for the time being. Land with a season's lease could grow wild flowers for a harvest of their seeds. Others, which would lie empty for a twenty year span, could grow trees for transplanting or coppicing on the spot. What a good idea.

Meantime, we had shovelled up our beefburgers and the families were ready to go. I waved goodbye and freewheeled down the main street to visit Roland Chaplain, a newcomer to Laurieston whom I had bumped into at the ceildh.

CHAPTER SEVEN

Roland Chaplain & Weather Watchers

I n 1985 Roland Chaplain set up a weather forecasting centre in his back garden, only 200 yards from Laurieston village hall. That winter, BBC Radio Scotland received so many complaints about inaccurate weather reports that they decided something had to be done. Geoff Sargiesson, Head of Light Entertainment, had struck on a novel way to restore confidence, and had asked Roland to call up the people who had complained, asking them to check the weather out of their window before the BBC wrote its forecast. They did it three times a day, live on air, as an interlude between the shows.

From these on-the-spot observers, Weather Watchers grew into a workers' co-operative supplying forecasts to, among others, *The Scotsman* newspaper and the Press Association. Their network expanded to 2,000 watchers, from mountain rescue leaders to the handicapped, from shepherds to all-night garage attendants. Roland had even taken on a marketing agency to sell the idea of a personal weather service for business decisions like the best harvest time and where to site your next road delivery depot.

Weather Watchers employed four full-timers and sprouted a purpose-built office in Roland's back garden. This fulfilled one hope to create jobs in the immediate area. Support staff had been drafted in from Laurieston village. Several at the hall had snapped up the chance for interesting part-time work. Also, the high-tech Apple Mac computer equipment needed to draw the forecast charts had spawned a desk-top publishing company that charged less to charities.

Work began at five o'clock every morning. Roland had to get up and call round various watchers to check the state of frost, wind or snow. Gradually he could assess the day's 'now-cast', as they called it, to prepare the forecast. He called John Wink, a shepherd who lived beside the A96 road and Bill Bain, a mountain rescuer at Glenshee ski station. Often he phoned people who were bedridden or handicapped and did not get the chance to chat much to someone from the outside world. Weather Watchers prided itself on the personal contact.

When I dropped in, the hive of activity had stopped buzzing for the day. A satellite picture had just come across the wires and did not change the outlook for *The Scotsman's* map, so it was touched up and phoned through. Thunder might hound me to Penrith next day, they predicted. Neil, the computer expert, showed me round—not the barometers and rain gauges or the offices, but a tour of the vast computer software. Weather Watchers even linked to GreenNet. Neil called up. On the screen in front of us there was access to a global computer network linked to over 100,000 Green activists.

They used the system to communicate with weathermen and women in California, Massachusetts and Sweden to talk about their problems and their finds. Neil had only recently discovered one of the regulars in the 'climate conference' was a world expert. He was most embarrassed at how simple his questions must have seemed. 'But it's part of the goodwill involved,' he said. 'I would not have been able to query him face to face but on GreenNet you have discussion regardless of race, sex or background. You can even talk with the Russian military, who are apparently quite *compos mentis*.'

GreenNet's publicity said that on a typical day an Irish campaigner added information on the latest submarine accident, a journalist in West Germany researched developments in US windpower technology, a solidarity group in Bath updated themselves on events in Central America (in English or Spanish) and sent a message to the project in Nicaragua which they supported. The list went on. After all they were trying to sell the service, which costs no more than the price of a local telephone call plus 6p.

To Roland, the work of GreenNet was highly significant. It had grasped the most advanced technology in pursuit of peace and intellectual liberation; people talked and developed their ideas constructively but also opened their eyes, literally, to a world of opportunities. Roland was well qualified to discuss the project, having spent several years as head of the Future Studies Centre in Leeds until 1985. (Future studies must not be confused with futurology, which predicts the next trend in lifestyles, the next colour in fashion.) Roland's operation had been run by volunteers on the dole. They distilled the ideas of some 200 radical groups and thinkers across Britain to bring out a quarterly newsletter. Of paramount importance was the cross-fertilisation of issues and arguments bandied about by separate people to effect change. People who would never have met were brought together on paper, then later bumped into each other at a conference to work out their differences.

Roland had explained something of the motives behind his idea outside the Laurie Arms during the ceilidh. He had been a more conventional academic when he was appointed head of Birmingham University's weather observatory.

'But then I was faced with a decision between approaching work in a way I considered unethical or making a stand on what I thought was most useful. It occurred at a number of levels. Was weather forecasting a science or an art? Where did it then fit in as a communication to help people make better judgements? Also I had to make the decision in reality because I was in charge.

'I was sacked when I made a stand against my employer's position. That really threw me into the 1968 movement. I found a lot of what students were saying or feeling at the time was the reality of what they were up against: existing value structures prevented progress.'

In this heady era Roland went to a world futures conference where he met open-minded thinkers with a divergence of ideas and lifestyles. 'From then on I felt it was important to do more networking between people,' he told me, 'between others who knew nothing about that sort of conference or couldn't afford to travel around the world to them. Many were doing very important and relevant things.'

So Roland had set up his Future Studies Centre and claimed Social Security. He had also inspired enough interest to offer a host of volunteers stimulating jobs (one former helper, Roland was proud to recollect, went on to take an upper second-class degree at Bradford University). 'I simply saw it as taking very low pay for a valuable job. My bad luck for doing so,' he quipped with a twinkle in his eye. It was sufficiently satisfying for him not to feel the need to spend money on other things. Having balanced the work-a-day world with an all-consuming intellectual experience, money no longer mattered.

It dawned on me that Roland was somewhat like Enoch Powell in his complexity of mind and yet clarity of expression. I asked if there were any times when he had put his own intellectual angle on the arguments and ideas he was collecting for the centre, or whether he wanted to be objective.

'There were times when I felt strongly on particular issues. Certainly, I remember one leading article on the peace movement. I wrote saying too much of pacificism was simply anti-weapon, anti-'the machinery of war', anti-army and not enough was concerned with offering positive alternatives to make it more attractive for people to be peaceful or to compete constructively rather destructively.'

Things tailed off as the sixties momentum was lost and the Futures Studies Centre folded. Roland felt that it had, at least, secured a place in the growth of contemporary ideas. Where other people plunged into their communes and their own things with heads down trying to survive, Future Studies had kept burning a torch of information and communication.

'I used the word "catalyst" quite often,' Roland said. 'I always saw our work as something that should not be fossilised into any structure but should draw people together who wanted to focus the ideas of a certain point in time.'

I wanted to explore this philosophy of fluidity. It seemed to match the reality behind my thoughts on the dynamism and plurality of Green philosophy, though I had not spent so much time trying to articulate it. Roland explained the genesis of his own ideas from the days when he was a student of both theology and meteorology.

'I started questioning the churches and above all the theological establishment at the time. Particularly, I focused on the gap between fossilising concepts in static language and the reality of the spiritual experience which lies behind a lot of the literature. The problem of language was that to define something, it pinned it down.

'Language evolved at a time when change was very slow and it was mainly designed to communicate what 'is' rather than the process of change and happening. Mathematics is capable of it, but language, other than poetry, is devoid of the subtlety that explores the realm of change and feelings. So, going back to Teilhard de Chardin on evolution, I'm asking: where are we going in the future? What's the meaning of all this we are trying to discover about the origin of conscious life on the planet? Is our traditional theology appropriate to cope with the phenomenon of man at the present time?'

Roland had recently gained independent financial security because his ninety-year-old mother had sold her Bournemouth house at the peak of the 1980s property boom prices. She had moved to live with him in Kirkcudbrightshire as bright as ever, still capable of argument. I wondered whether Weather Watchers could contain Roland's spiritual and philosophical insights in its daily business.

'One has got to maintain viability. We have to fit in time deadlines. You can't have philosophical discussions if your copy ought to be at the Press Association by five. It's a matter of balancing these things. If you know you're going to do this, it's fine. You find other times to talk about community businesses and co-operatives and the structures that enable people to have more control over their work in every sense. That means quality of work, the relationship with clients, the choice of who you're going to work for. Our discussions here are about all these.'

Roland made a great deal of sense in his simple language, yet charged it up with long pauses before giving an answer—and glaring right through me. I felt my questions were rather opaque and lightweight when he replied with his lugubrious, Gielgud tone. Before I arrived, I had underestimated how profoundly he viewed his work and life.

'There are other sides to my life. For three years I've spent an hour or so a day building a wall out into the loch for a swimmer's beach. I think it's important to do something practical. That's what the whole ecological movement is all about: there is strength in diversity. We need to be 'doing it' personally as well as working with diverse ideas.

'I think it's very dangerous always to be with like-minded people. You've got to throw yourself into the real world where people eat things differently, have different lifestyles—and not be arrogant and say 'ours is better', but always be open. Things may not be the same as you think or you've read that they should be. That's one of the dangers of reading too much and learning a way.'

We talked of the privilege education had given, and of its tendency to brainwash; of how the pursuit of money seemed to sap intellectual curiosity. Since Roland seemed to have such a clear point of view on both complex and simple ideas, I went for a big question. Why did he think money had taken such a strong hold on society?

'It's a competitive thing. People have been so much encouraged to compete for more of it. And it gives status. And they are insecure and want to be recognised. And it gives power. And you take more women out. And you can splurge it about. You assume that is what is expected of you.

'There are just so many misconceptions given to young people in education. This was part of my battle with the education establishment in the 1960s. They were just giving the wrong values. It was all about competing to get high academic qualifications so you could earn as much money as possible and get to the top. They had lost the ideals of knowledge for its own intrinsic value.'

I presumed that he would also have an interesting insight into the concepts of change and transition, considering his expertise in forecasting. I asked him what could transform people's acquisitive natures.

'You must get people to start asking the right questions, like "What do I really want out of life?" This is happening: more are questioning whether they should use conventional high-tech medicine; whether they really need lots of material things pushed at them by the advertisers, as opposed to liberating things, like fridges and dishwashers, that free them to have more time. You've got to be selective and think: "What do I need and what can I share." The stupidity of it is that people own things all their neighbours have, but only use a few hours a year. I think people need to talk more to share more.

'It's something you learn in the country. I felt that when I walked around Iceland about twenty years ago. That was my first contact with isolated communities and the way in which the people co-operated. Because there, it was a matter of survival. If you didn't lend to your neighbour, they wouldn't lend to you when you were in trouble. There's nothing better than necessity for co-operation. It's when we have too much, too much affluence, that we have problems.'

The fact that Roland was so close to Laurieston Hall meant he had a close community to bounce such ideas off, but he also practised what he preached about sustaining diversity and made sure he had contact outside the circle of like minds.

'With my wife's illness and eventual death, I'm quite sure I would have cracked up in a city situation,' Roland said. 'There's a different form of support here. A greater level of trust, essentially. People have been fantastic to me. So I hope, when the situations are reversed, I can pass something on that I have learnt this past year.'

'Is there a sense of community?' I asked.

'Yes, that's it. Because it's small, people care much more about others, like the elderly, when they are really in need. Obviously, as in all villages, we have the neighbourly squabbles, the backbiting, but when it comes to the crunch people care a great deal about each other.'

Roland mentioned that he had been involved in a circle of co-counsellors, mostly from the Hall. The discipline involved a strict set of rules for one-to-one therapy. If one of the circle wanted to unburden an anxiety, another would sympathetically listen and offer advice. This co-counselling had fallen by the wayside, but the innovations at Weather Watchers had not stopped. The latest was the massage exchange. Late morning staff would shoulder-massage their colleagues who had started the early calls, to soothe the stress of the hot seat job. 'Of course, all the big businesses and the City are doing it now with their own on-site masseurs,' Roland joked. 'It's funny how alternative things are spreading down into the mainstream, recognised as having a hard economic value—it means you've not got your expensive executives breaking down or in pain all day.'

With Galloway and Dumfries on the doorstep, and all that fresh air, there was a lot of free nature therapy. Pity those commuters with hundred-mile daily rounds.

'What do friends in the South think of your move?'

'They'd hate to break with the security of their discomfort,' Roland said. The sentence bounded out and came to a dead stop. 'That's a good one. You can quote me on that. People are afraid to do something new, however attractive it is. Lots say: "Oh I'd love to do that." We get interviewees who say the same, then they don't take up our job offers. They stick with what is familiar; the security of a pension, the structure of the Met Office, conventional things. We've got a huge legacy of insecurity that traps a lot of people into very boring, repetitive, uncreative jobs and lifestyles. I just hope our success can show people what can be done. That matters to me a lot.'

CHAPTER EIGHT

Nick and Ana Jones, The Watermill

I had an early start, having spent one night in the Goathouse, a converted wing of the Hall at Laurieston. I had left the dormitory where I had spent the rest of my visit to make way for sixty gay men up for a week's holiday away from the glare of a homophobic world. Mist still hugged the ground, chilling the first miles of my ride and, in my haste to warm up, I forgot my promise to pick a sprig of the four-leafed clover which was said to grow in the kitchen garden.

At eleven I passed Ruthwell on the Solway Firth flats by Annan. Generations there have guarded an Anglo-Saxon stone cross inscribed with poetry I'd learnt at university. A sign advertised the Annan Savings Bank Museum, which commemorated the work of a local vicar who, in Victorian times, had been custodian of the cross, and had who set up a local Annan Bank.

The bank had closed but I imagined it was like an early version of the Totnes Green Pound, which Pat Fleming promised I would find at my final destination. In Annan, the enterprise had been superseded by mega-bank muscle, but history had not been totally eradicated. The Annan Savings Bank offices stood still, empty and dusty.

I pedalled on, past Chapel Cross power station, taking the fellside road to Little Salkeld in Cumbria's Eden Valley. It was a picturesque scene. Rolling hill farms girdled the land in tight belts of drystone wall. The evening rumbled to the tune of hay gatherers' machines. At dusk, I arrived at the Watermill, run for fifteen years by Nick and Ana Jones. The house, mill and animal sties lay only yards off the roadside and were signposted 'No longer open to the public'. There was peace and quiet about the place. The cat prowled to its heart's content, hens strutted and only the crunch of gravel interrupted the silence, as I wheeled KP Crisps to the front door. Kittie and Harvest, the Joneses' two daughters, sat masked by blossoming roses in the last sun and drew pictures. I was taken aback at how grown-up they were and how chatty. But then they had lived practically all their lives with tourists round about them. We talked about Lindisfarne, where they went for family holidays. Harvest was an avid and knowledgeable birdwatcher.

Nick greeted me on the threshold. A tall man of forty going on thirty, he stooped through the open front door and offered apple juice. There was a Dickensian quality in his fresh looks, somehow appropriate for a miller. He had long dark hair, pebble spectacles, a striped shirt and beige flannels rolled up to show that he wore no socks in his suede boots. Of course, there was a dusting of flour. Nick had the formality of a period actor and the Georgian house, his back-drop, was a refined and lived-in film set. I sipped my drink outstretched on the gravel and recovered from the day's hot, hard labour before supper. Harvest, in a singing Cumbrian brogue quite unlike her father's Oxbridge vowels, told me Lindisfarne was her favourite holiday destination. She would not take the family anywhere else. Kittie drew fruit and portraits on the sketchpad.

In 1974, just married, Ana and Nick had been on the look-out for a smallholding on which to practise self-sufficiency. John Seymour's book on the subject had inspired them with the dream of a cottage, a goat and seven hens. Nick had run Rosehill, a theatre near Whitehaven, Cumbria, known as the Glyndebourne of the North. Ana was an artist by training and had practised in London. They had seen the mill and knew it needed their young energy. 'It was important to keep it going,' they felt. It had been run for a hundred years in another family's hands but Jack and Mary Atkinson, the last generation, had decided that they could no longer keep up the hard slog. They had sold the mill in good order with a seven-acre smallholding at the back, relieved to see the tradition live on into the future. The Joneses were willing, though not very able.

'It was one thing to dream of a more practical, down to earth lifestyle, quite another to live it,' Nick said.

'A mill was certainly not in the deal,' Ana added. 'Nevertheless, in the beginning, we were fanatics. We did everything and somehow managed to succeed. I don't know how. There were sheep, goats, vegetables, even regular home-made ginger beer, bread and cheese. Don't forget we also milled our own organic flour, ran the café and saw to the tourists.'

To buy the mill in the first place they had needed a grant. The deal with the tourist board had obliged them to show visitors round for ten years. It sounded like a prison sentence, but they had no choice. Each year, 10,000 visitors trooped through their garden and the old watermill works, then had tea and scones in the café. 'People didn't realise that it was like having three houses, with three floors to sweep for starters.'

Still, they made a conscious effort to radicalise the café. They banned all sugar to set visitors' thoughts on diet and food and where it all came from. Nigel Wild, back in Newcastle, would have applauded the idea. He had mentioned the injustices of the sugar industry when I was at the bakery: sugar workers were slaves to the estate owners who sectioned off their land. Nigel had recommended that I stop at Little Salkeld on my way south between Laurieston and Yorkshire. He had learnt his trade at the village bakery in Melmerby, just a couple of miles from Little Salkeld, and his flour was Watermill flour, delivered from Cumbria in beautiful brown paper sacks and stacked in the cellar.

By coincidence, Nick had been in Newcastle the day of my arrival, on a delivery. He had picked up enough goodwill pizza to feed the family and Dennis, an organic gardener who had hitched a ride back to Salkeld. Dennis organised the Joneses' garden and came over from Durham when an extra pair of hands were called for to make hay. Over our pizza I explained my mission, and talked about a few of the people I'd already visited. Dennis recounted the labour of love he had performed at the Henry Doubleday Research Association in Ryton Gardens, Coventry, which I had lined up for a visit. He had sown peas to exact spacings and measured them up with the precision of a micro-engineer as they grew—not the freehand gardening he favoured. Dennis had also spent some time working for Tools for Self-Reliance, a charity I had encountered at Glastonbury. He encouraged me to visit them. Kitty read us poems about witches. Harvest argued about animal testing and vivisection. We all answered a quiz to test our Green commitment, which Dennis won because he had no car and could not drive.

The kitchen table where we ate stood in the bay window at the back of the house. To my right, I could see the black iron waterwheel, stopped for the night. The mill race dropped directly below where I sat. I listened to the soothing sound of running water, as the blue sky turned to a deep glow behind the black staves of tree branches.

I asked the Joneses if they were resented as incomers: they told me it had never been a problem. Besides, they had given worthwhile jobs to all sorts of people by keeping the concern going. There had been the café and the mill itself. The brown paper sacks had even spawned a cottage industry which employed a retired friend—learning the printing craft had been an invigorating experience for him.

'There has always been a tolerance of eccentrics in Cumberland,' said Ana, 'especially with all these small family farms around. They're used to it. The people are all quite in-bred, but the families have managed to survive through thick and thin with their own tiny units. We were just another part of that, and have quite a lot of friends doing similar things.'

Nick had always been driven by a desire to integrate with the community. He wanted to put down roots because his own upbringing, with a father constantly abroad with the diplomatic and cultural establishment, cut him off in a rarified, transient atmosphere. Love for the mill and for Cumbria had gone so deep that they would never fully give it up. It had given them a livelihood and a standing in the area. 'I can quite confidently say we have friends of all ages and across all classes in the village and the area,' he said.

'Yes,' Ana added. 'One night we might have dinner with a very old couple with an estate on the hills, the next we are in some village hall.'

'The only level of society I miss, though,' added Nick, 'is the university class. There is no higher education to speak of within miles of us here. This is the worst-served part of the country for that. There is no university anywhere in Cumbria and I think we miss out.'

Even so, it sounded to me as though the Joneses led a very full and fulfilling life. I was envious, too, of the whole beauty of the enterprise. The antique furniture in the house had been chosen with an eye for period charm and matched the warm feel of the stone walls. There was oak and pine throughout. Up in the mill, besides the rickety, steep stairs and rough carved, cog engineering, stood a high, clerk's desk. I expected to see Nick stand with a quill pen to count out the orders as they were craned down to the wagons below. In the hall was hung a prized family possession: an unforgettable, bright painting by Winifred Nicholson. 'We were lucky enough to get to know her in her old age because she lived near,' Nick told me, with pleasure that compensated for the hardship it must have taken to acquire the picture.

I imagined Nick's Cambridge University friends would visit in their hordes, that family and friends in the south would always be popping in. After all, this self-sufficiency gamble of the 1970s had become rather fashionable by the late 1980s. But university contemporaries no longer called. They had not asked Nick and Ana down South for ages. 'We haven't been to one of their parties for a long time,' said Ana. 'And they stay away from here. People find it hard that you can size them up in a minute—their lifestyle, their thinking. And we don't fit into the scheme of things there.'

So things looked quite different fifteen years on, life honed by Harvest and Kittie, aged twelve and seven (seven and a half, actually—I was corrected). There were no more shocks and surprises brought by heavy rain or stray visitors, no need to scrimp and save (though few extravagances either). The miller and his wife now had time to concentrate on going back to some community arts and crafts. Nick had become involved in running a local arts trust. Ana had started a new business, weaving the vegetable dyed wool sheared from her unusual sheep.

After the washing up, Nick took me to survey his land. The seven acres were a narrow slither up the river bank. A motley flock of sheep and goats, in breeding pairs of various delicate species, filled the first fields. Up high, with magnificent views, the hayfield had been cut, thanks to Dennis. The steep bank to reach it was planted with trees, now nearly a decade old.

It seemed appropriate to think of what Nick had written for *Resurgence* in 1985:

'It is not just a matter of living and working at home, of producing something of value in the best possible way, of living off and with the earth and its creatures. It is a matter of doing and thinking all these things in the right spirit and communicating that vision through our way of life, our products, our place.'

I asked Nick what his views were on such evangelism in 1990. He no longer had the headache of 10,000 afternoon teas in the radical café. Instead, he had to face the fresh rebellions of his daughters who hated organic farm fresh produce, like the goats' milk they lived off. 'Evangelising your ideas is a disaster,' he replied.

Some calm was restored when the public's glare went elsewhere. 'I am now more balanced... centred... grounded...' Nick told me, slipping into Buddhist terms. His occasional retreats at the nearby Throsselhole Zen monastery had influenced his phrases. 'The mill was a journey for others when we were inundated with visitors. We were permanent crew members stuck with the boat, they were the travellers. Now we want to keep the boat but sail it for ourselves. Without it you completely lose the means to choose where to go.'

Hard labour had recently been shifted to other staff. And neither of them actually milled the flour any longer, or worked in the print shop to stamp their distinctive sacks. The relief of five years without visitors had given them a new lease of life and now a trickle of 'WWOOFers' come regularly for a 'Working Weekend On an Organic Farm' organised through the Soil Association. It was halfway between having travellers and having the ship to themselves.

We walked back to the house past the flood gates which diverted the stream into the mill race. The height at which you could drop a good load of water on to the wheel was the crucial factor for well-powered milling. Because of the narrow incline, Nick's race was 150 yards long before it had gained sufficiently to drop back into the stream. It had been dug over a hundred years before, but every so often needed a good digging to cut back the weeds and speed up the flow. I signed up for a morning shift after breakfast next day, when the family were off in different directions: Nick and Kittie to the community arts paper-making course, Ana and Harvest to the east coast for a birdwatchers' meet.

Over coffee, Ana talked animatedly about the 'WWOOFers'. A young couple drove over for their last weekend in a live-in van and Ana looked back nostalgically to the time she and Nick had the same spirit for hardship and an energy for the hours of cheese-making. Nick silently pondered. Ana concentrated on the new-spun wool she was busy refining. She looked younger than her years, which she put down to a good, natural diet. The air gushed through the open windows and brought with it the beat of water sluicing down the race. This meditative hum powered the whole enterprise. Nick picked up the thread:

'I think we see our next port of call as simplifying our lives with self-sufficiency in a material and a psychological sense, because being Green is about psychology, among other things. It is the ability to come to terms with yourself, the old thing of knowing yourself so that you can live with that before you can start to change and before you can work on the external things.'

Nick wanted to go on learning new things and developing as a human being. Was such sentiment merely an educated person's luxury? Did development in fresh directions result in more consumption and spending? 'We do feel we are custodians of this place,' he stressed again, 'but we are torn between sharing it in whatever way is best and cutting ourselves off to do our own thing.'

The sharing was easily done and most tiring, but the balance between simplification and unselfishness was harder to achieve.

Neither Ana nor Nick could see a simple solution yet. Like the faithful, they were prepared to sit and wait for the right inspiration.'Anything's possible,' Nick said. 'When people say it isn't, they mean: "I don't want you to try" or "I wouldn't like to do that." I always think it's a challenge to prove them wrong.'

CHAPTER NINE

Bob Lowman & The Ecology Building Society

S outh again. The Carlisle to Settle train rocked me into a doze. I had hardly slept the night before, though not for lack of comfort—I had spent it on a First World War straw palliasse in the whitewashed barn loft. The dapper cockerel in the henhouse next door had decided to crow from five in the morning on.

The train ride was fantastic. The scenery was so spectacular as a tourist attraction that the route had remained sufficiently viable, even through sparsely populated moorland, to stave off closure. At every stop tourists packed into carriages with regulars on their way to Leeds or Skipton for shopping.

It was a day out on the move; not stuck behind a steering wheel, cramped or gasping for air as in a car. A breeze whistled through, and all the family could move from side to side as the scenery went by: houses and ditches, heather and fells.

We stopped at Kirkby Stephen's West station, which suggested the little town had more than one originally, an achievement for its size. A children's birthday party climbed aboard. Five little blonde girls squeezed on to the three-seater bench in front of me, their bunches and pony-tails bobbing as they sang in Cumbrian whispers. The birthday girl's mum carried towels and swimsuits and tensed with their shrieks. Train travel was convenient all the same. She could keep a watchful eye on all of them without needing to concentrate on the road as well.

I thought of equally interesting country routes home in Northumberland which could be reopened to keep the villages breathing. But twenty-first century development on that scale would need imaginative and large-scale vision. I got off at Skipton. It was only half an hour by bicycle to Crosshills.

Bob Lowman, the manager of the Ecology Building Society, stood up from behind his desk to shake hands. They looked a safe pair in which to hold the society's £6 million assets. He was dressed in a grey suit and striped shirt, probably Marks and Spencer. Crosshills folk who saw the company as a hippie enterprise had misjudged it. The business had been tidied up over the years. The only ornament in Mr Lowman's office, in a converted terraced house, appeared to be a photo of his granddaughter in party frock eagerly clutching a coffee-table version of *Small is Beautiful*. *The Times* was folded in the in-tray. Mr Lowman was definitely not a hippie. He was also not a rash man. He prefaced anything bordering on gossip with a 'Don't quote me on this'. When he speculated, he asked not to be quoted. If I pressed, he looked up the facts. Peering skyward, he tried to remember his curriculum vitae. He had spent thirteen years in banking before a switch to building societies for seventeen, which brought him from London to Crosshills and to the Ecology Building Society. He played down two 'excursions' into charity work. First he had run a community centre as a Baptist pastor in inner city Battersea, but after two years the evangelism had got to him. He didn't think pushing church invitations to 6,000 people in high-rise flats had the personal touch. He had also organised the Children's Society shops in Yorkshire. He had escaped London after his children grew up, to slacken the pace. 'It was an especially interesting management experience at the Children's Society,' he told me, 'because I was training volunteers rather than people you paid to work.'

He had also come north because it was the place of his fondest working memories. He had manned the counter at the Halifax Building Society in Halifax nearly twenty years earlier. Mums would come every week from the nearby Dales villages with savings books for each of the family to have them stamped. There was a true grit in their daily life, a personal touch in his.

Mr Lowman sat opposite me and ate his sandwich lunch from carefully packed, clear plastic bags. I could not imagine him in tears, but he said he had nearly cried when he saw the Ecology's promotional video. The film described how the Ecology saved the small village of Nenthead, Cumbria. Loans were made to renovate a row of run-down miners' cottages and to save them from outside developers. They were transformed into low-cost housing for the locals. No one was forced to leave their native district, and the social ecology stayed a healthy mix.

'To save a house from becoming a second home or from rotting is a good enough ecological reason for a loan,' Bob Lowman said, mustering up spirit and laying down some of the criteria by which the society considers mortgage offers.

The Nenthead village school had staved off closure as a result. The local post office kept its doors open. The whole village was eternally grateful that their lives were not ruined by long journeys. And it was all down to a good, honest loan.

The business of the Ecology was simple but principled. The society only lent money if a housebuyer respected ecology, defined as 'the totality of man and environment' or 'the link between the natural and social sciences'. More specifically their rules stated that someone could qualify for a loan if:

a) they saved resources in renovation or energy efficiency;
b) they promoted self-sufficiency for individuals or the community;
c) they kept nature and land in good heart.

'We have made 200 loans to date and they're all special,' Mr Lowman said with executive pride. One, for example, was taken out by three women who bought their village store when it was about to close. If it had, they and the rest of the villagers would have had to travel five miles to the nearest shop. So the women shared the business and could offer one of their widower fathers the upstairs flat. A community institution had been saved and an old man housed.

Ecology loans span Britain in every shape and form. There are renovated steadings for therapeutic retreats and health food shops with flats above. At the remote town of Alston, Cumbria, there was a group of eight Ecology mortgage holders and the lending brought terrific community spirit into the area. Two of the borrowers started up a business called Green Ark which made wholefoods and homeopathic herb remedies for animals. These had sold well and the business had grown to employ four people.

Mr Lowman took a telephone call. A gentleman had begged money from the high street building societies for a house on moorland hundreds of yards from the road, but none had offered him what he needed. It sounded an idyllic spot, but the cottage needed attention. The buyer was prepared to pay a sum of money which nobody, not even Mr Lowman, thought realistic. The borrower had been turned down without any explanation by the major societies. The Ecology was at least prepared to talk through the criteria that they take into consideration, and suggested ways that he might be able to satisfy them.

Another interesting project, Mr Lowman continued, once his caller was despatched with a few options to think about, came out of the Lees Stables. The commune took an Ecology loan to renovate some old farm buildings in Berwickshire. Two families there put together *New Cyclist* magazine, developed all sorts of weird and wonderful cycling contraptions and farmed a smallholding. This kind of smallholder made up the core of borrowers. The Ecology also lends money on terraced, back-to-back houses because they are potentially very energy-efficient and could save resources by cutting commuter mileage for their owners.

On paper an Ecology mortgage costs more than most, but price is not necessarily a deciding factor, Mr Lowman told me. On the 'unusual' houses considered, say a wooden cottage, the choice is often between an Ecology loan or no home and the main reason for the higher price is a legal one, rather than one of small-scale finance. New laws require every building society to hold cash reserves big enough to cope with large withdrawals of borrowers' money. The big established competitors had funds on hand but the fledgling Ecology has had to build theirs up.

'We're a value-led business not a profit-led one,' Mr Lowman told me, quick with the words of The Body Shop's Anita Roddick. 'We call it recycled money.' A 'Green Money' pamphlet I read when Mr Lowman was on the phone made the point: 'Money is neutral— it is what is done with it that gives it meaning and power.' Mr Lowman did not like the word 'profit' used in the building society context. He explained his views as follows:

In the early days, a group of people would get together to spread the burden of building their own homes. They would pool money and, as homes were finished, would draw lots to see who moved in. Then they built the next house and, with the last savers housed, the society folded. These 'mutuals' and 'friendly societies' grew and membership changed on a rolling basis. 'Permanent' societies formed, traditionally composed of friends and neighbours. Today's biggest societies retain the names of the towns from which they grew—Halifax, Leeds, Bradford and Bingley, Leicester, etc. The principle of a money cycle remained. Some lent, some borrowed. The society took a cut to cover costs. Owned by its members, it acted for their best interests—literally. That was how it worked until Abbey National transferred ownership to shareholders, and became a financial institution like any other.

Building societies thrived in Yorkshire at places built around the textile industry in the late Victorian era. Thrift and self-help ideals had much to do with it, Mr Lowman reckoned. The liberalism of landowners must have been a deciding factor, too, I presumed. For, if the owners refused to sell their land, people remained dependent on them. Factories might have lined my route through Northumberland and Dumfries, but the landlords there had kept it for agriculture. Marginal land, like that in the Dales, went cheap to the pioneers and so gave way to the Victorian boom. Perhaps there was still something today for the smallholders of Wales, western Ireland and the Highlands. They had taken over where space and price allowed and one day their developments might be monuments in their own context—the 1990s equivalents of dark satanic mills.

Mr Lowman reminded me that building societies had flourished around other communities too, those of shared values: the Teachers', the Temperance, the Catholic Building Society. The Temperance lasted until 1974 when drink was found in the boardroom and the members revolted. The divide between members and management, which later split Abbey National, had already begun. Sir Cyril Black, at the time Mr Lowman's MP and 'a great temperance man', led the fight. But values lost out to the Temperance board, who renamed their business the Gateway.

The Ecology moved into a field of stiff competition, beginning the same year the Temperance closed. A solicitor, an architect, an accountant and a printer championed the cause of a smallholder who was refused a mortgage for seemingly political reasons rather than for sound economic ones. Their brainstorming session came up with the idea that they set up their own building society.

Ten people each put up £500, then the minimum amount required to register under the Building and Friendly Societies Acts, which guaranteed solvency through a system of mutual responsibility. The new-born Ecology ran from the solicitor's office, then took two rooms above a flower shop in Crosshills' main street.

With growth came responsibility. The idealists who began the enterprise gave way to the realists who wanted to combine competitive rates for clients with ecological values. Roy Pickard joined the Ecology on a part-time basis. In 1985 Bob Lowman was taken on by the board of directors as full-time manager and new offices were acquired in 1989. He was later joined by another part-timer, John Cooke, who together with Roy Pickard brought a wealth of experience of the building society movement. True to building society and ecological tradition, the Ecology remained local in operation. In fact, half the staff were educated at South Craven School in Crosshills. Two YTS trainees, Joanne and Gillian, were recruited with guaranteed jobs and there have been other knock-on effects in the town. The local stationer, for example, changed some of his stock over to recycled paper to supply the Ecology's order. 'We weren't looking for growth *per se*, but demand for our services meant we needed it,' Mr Lowman said, surprised that things had gone so well.

His expertise came in useful because a spate of financial deregulation had transformed the game. Eight thick volumes of regulations groaned on the mantelpiece waiting to be ploughed through, and that was only last year's new legislation. Lucky he was a details man, and up to the task. The grip of bureaucrats made his blood visibly boil. To flex his muscles, he launched a campaign to change the legal definition of smallholdings amid the labyrinth of the new laws. 'Somehow we've also got political patrons from each party to whom we send copies of our minutes. They don't usually take an interest, but sometimes they pick up on things.'

Another man from the ranks of the great and the good has taken an interest and spoken at the Ecology's AGM. 'Jonathon,' Mr Lowman said, 'has been a great supporter, and came to our Sheffield conference in 1988.' He meant Jonathon Porritt, of course. Mr Lowman was on first name terms with the green guru himself, so I thought he might shed some light on the green yuppie question: did you have to be rich to be able to afford to be green?

The Ecology had given fewer 90 per cent mortgages than any other society and, on average, lent only 50 per cent of the cash needed to buy a property. The largest loan they had made was £80,000 for a wholefood business with flat above. The average was a £20,000 loan shared by three or four people on one property—but many of those were canny types who bought cheaply and renovated beautifully.

'But yes, we have lent to a TV producer,' Mr Lowman said. 'And I saw the hyphenated name of one of our borrowers the other day. He had a letter in *The Times*.

'So no cash flow problems at the economic down-turn?' I asked.

'We work on a friendly basis. People usually get on to us pretty quickly and we have less in arrears then the national average. I've talked to a vineyard owner who is behind on payments, but he has 14,000 gallons of British organic wine ready to bottle, so he'll get sorted out.'

I opened a savings account, to transfer my Abbey National pittance and lose 20p interest a year. All Ecology transactions went through the post or giro transfer. I pondered whether this was of ecological benefit and could not decide either way.

Outside, I changed from my respectable trousers into shorts for the road, and was gone. I took the road for Bolton Abbey, to stay with Louise Woodstone and her four-year-old son Tom, whom I had met at Laurieston. They lived in nearby Beamsley, a village almost wholly owned by the Duke of Devonshire, the area secured for natural beauty and grouse shooting, not peopled, mined or milled.

A wrong turning took me up past the Abbey grounds themselves, where a million day-trippers stroll each year. The estate had laid down expensive solid walkways to stop the erosion. All tourists had to tramp the grounds on set paths. What price access? Where were the mills and factories and homes on the Duke's estate? I turned in search of Beamsley village and discovered my panniers hung unzipped behind me. All my notes were gone. They had dropped out since my picnic lunch on the hilltop. In panic, I retraced my way with Louise and Tom in Louise's car and found my black bag of notes neatly squashed into the tarmac outside a pub three miles back. Two men sat on the grass outside the pub, empty frothy pints in hand. They caught my relieved look. 'We thought it looked important,' one agreed with the other, but neither moved.

CHAPTER TEN

Suma Wholefoods Co-operative

N ext day I baked and strained on the hills to Halifax. Suma Wholefoods run their nationwide distribution from a massive warehouse there. I free-wheeled into the Dean Clough Trading Estate. It was lined with cars and trucks to support the £7 million-a-year business. About a third of the Suma staff sat at a meeting outside. Gill Hilton, who was responsible for marketing, stood to greet me and apologised about the gathering. It was one of the three grassroots meetings each week for everyone to have their say on issues of work. Meetings, I had discovered, were the bane of co-operative living, but less so at Suma.

Gill took me up to the office kitchen where I could choose from, barley cup, teas, coffee, juices, milks and soya milks. Compensation for the communal movement's meetings and 'structures' undoubtedly came through food. Always and everywhere, apart from Glastonbury's clean-up canteen and the sorry night I camped on Bruce Marshall's worm-infested land, my stomach was contentedly filled.

Bob was cooking the lunch given to all the workforce. There were nut fritters, mushrooms, rice, salad and lip-smacking sauce, followed by fruit salad and yoghourt. But there's no such thing as a free lunch. 'You couldn't just give us a hand with the fruit salad?' said the look on Bob's face as I leant back to sip my tea. I volunteered for washing up and grape de-pipping and we chatted above the Archers and the clanging shambles of pots, pans and fruit. Bob was not a Suma member. He was the organic vegetable delivery man and came every day via Nottingham on a circuit of small customers. He was regularly roped in for chef duty, he said with glee—only a short swallow away from full co-operative membership.

We talked about the sinister goings-on at corn circles; about a policeman friend in nearby Todmorden who said he had been picked up by aliens in a UFO, and never contradicted himself under hypnosis and other sorts of scrutiny, including American TV. I asked about Hebden Bridge, where Bob lived. I'd heard it was the 'Totnes of the North' with a fabulous array of people. But, as yet, nobody had given me any specifics. Bob knew a rock musician's family he thought I ought to meet.

At lunchtime I sat in the yard. The women lay and sunbathed. The men played football. Gill Hilton talked me through the business. The workers' co-operative had begun in Leeds at the right place and the right time. They reckoned they were the very first wholefood shop in the North. It grew and began distributing to other fledgling shops. The co-op juggled with a cumbersome warehouse in Leeds before they came to Halifax and a purpose-built distribution hangar. They filled the ground floor expanse with 3,000 different 'lines' of goods for sale to food co-ops, shops and smaller wholesalers.

Suma employs over sixty people, each paid £10,000 basic full-time wage with a bonus for children. They are involved in the development of the Fair Trade mark with Richard Adams, to brand goods which are not just good for the consumer and the environment, but also good for the workforce.

'Who's in charge?' I asked naïvely. I should have learnt a more subtle line of questioning after visiting Laurieston. 'No one,' came the reply every time. No one would even admit that one group was perhaps less in charge than the others. Moreover, everyone was allowed a chance at any job they fancied, and the company would train them to do it. The formula had obviously worked so far, while so many other co-ops had failed.

It appeared to be a success because it was businesslike. Suma tried to beat the competition in service and price. I browsed through the cash and carry. There was box upon box of goodies. How could one ever be against consumption? I was back in childhood days when a box of forty-eight Mars bars and a tub of Welsh's Football Chums meant paradise, except that here they stocked thirty-six different types of soya milk, flour from the Watermill and dog meal from Green Ark in beautifully simple brown paper sacks with decorative labels. Cases of organic honey,

a connoisseur's collection of organic and herb tea, organic hazel-
nuts, walnuts, brazils so fresh the oil oozed from the sacks.

The chains had lopped the margins off 'Green' cleaning products
and stocked their own brand wholefoods, so Suma had changed
the game. Suma's latest push had been on the supply of a wider
range of organic produce than any supermarket, Jon Knight told
me in the marketing department.

The warehouse hummed with activity, accompanied by
Mancunian drones of The Smiths piped over the tannoy. Women
drove forklifts in dance manoeuvres to pick crates of flour and
stack them by the Hebden Watermilling Co-op's corner. This
separate company was contracted to weigh out and package ten
kilo sacks, five kilo bags and smaller ones, ready for the shelves of
the shops that Suma supplied.

Electric carts zoomed along the aisles, driven by pickers. Their
robotic ballet went from stack to shelf, shopping on a grand scale
for customers' orders. They ticked off: ten woks, seventy-two loo
rolls, twelve kilos Nicaraguan coffee, four kilos tofu. The order,
double checked by stock control, was wrapped in yards of cling
film plastic ready to dispatch as one big lump somewhere on the
Suma fleet of seven trucks.

In the office, four staff sat at desks strapped with state-of-the-
art telesales headsets. Around them, though, was the normal
office hotchpotch of old furniture and organised heaps of clutter.
There were desks for buyers, financial accountants, management
accountants, customer accountants, credit controllers. A good
quarter of all staff work on the administration side. Strict big-
business organisation kept Suma going.

Frank had spent six years in credit control chasing debts. He
was trained in-house like all the accountants. And the secret of
Suma's success according to Frank?

'You have your say, you get involved, you are happier, and
therefore you give more. There's no suspicion of a rip-off. Even in
the bad times you keep your head down. There's not the same
tendency to complain about things when you've chosen to be here,
when you are employee and employer. And if you don't like it you
make yourself redundant.

'In a straight business the decisions are not necessarily to do
with the people involved. Often they are to do with the profit
which can be drawn off for the managers themselves. So to talk of
our viability is to talk in different terms.'

Specifically for Suma this meant deadlock over the pay structure.
The necessary consensus could not be reached. One faction

wanted to keep the same wages for all, another wanted to recognise length of service. They argued that old hands deserved more because they had put in more and that experience needed reward. Others on the fringes had called for pay to be hourly or to follow the industry rate. Without a solution, you solved the problems yourself. Stay and argue it out or go. Frank had resolved that if the pay was going to be the same no matter what, he would treat the job as a nine-to-five business. Idealism did not mean you had to exploit yourself. Suma was life for some, but the company had reached a stage where it did not exploit its workers out of the need to survive.

'You've got to remember you're a human being as well as a member of the Suma co-op,' Jon Knight said, hinting at the intensity on the inside. He joined specifically to combine his retail experience with his interest in the co-op movement. In fifteen years he was only the eightieth worker to join the company and sixty were still there. No one in company history had been made redundant and a sacking was rare. 'Things work because we look after each other,' he said.

Matt Pinnell explained that Jon was one of a new breed of recruits who chose to take a pay cut in order that they could see how Suma administered its business structure. He could put it down to experience on his successful career journey in the 'straight' world. Matt himself had been unemployed and without qualifications when he got his first Suma job five years before. He was lucky to get training and rose to an accountancy post. 'Things have changed now and, in a sense, I would never be able to repeat my career here,' he said as I prepared to leave for Hebden Bridge. 'Now we only take on people with a good record of skills.'

Consolidation at Suma meant professionalism, as it did at the Ecology Building Society. It was a process of creating something more conservative out of radical roots in order to support financial livelihoods. Suma, Matt reckoned, survived because of the conservatism of a boss with sixty different heads to feed and satisfy. A change in structure needs a three-quarters majority, so as the business grew it never had wild changes of direction and everyone stuck to their truck-driving, accounting, cooking or selling.

As I left the warehouse I read the career opportunities notices. Offers included delivery for one morning a week, two days a week book-keeping, an afternoon on the checkout. Truly a chance to create a varied work-load for yourself. You would never be threatened with the same desk for forty years, as Bob Lowman had been when he was in charge of the Lambeth Building Society.

Another notice was about as categorical as the management

could get: 'Smoking in the warehouse is not open for discussion.'

Hebden Bridge, 'Totnes of the North', lost some of its attractiveness when Matt Pinnell whispered it was 'the place where old hippies go to die'.

I passed a clog factory museum beside a canal moored with Romany rose-painted barges, and my hopes for the town foundered. Anything distinctly and deeply Green, rather than respectably 'folk-craft-heritage' was well masked behind supermarkets and traffic. Hebden was just a dip in the landscape with cosy houses either side. Only the local art gallery looked promising. They were selling hand-printed copies of a ballad on the Ramblers' Associations' first mass trespass. I left, impatient, to climb over the hills to Gargrave where I was to stay with an old school-friend. I was depressed to find so little concrete evidence of a Greened town and felt it to be a failing on my part.

The intensity of my quest was forgotten as I relaxed with my friend Richard Emmott. The chat changed from the Green agenda and culture change to the tennis club, the family and the furniture business, but I could not resist one up-date. A couple of years before, when Rick's father was a company director in the timber trade, I had asked him what impact tropical wood campaigners had had on his business. They hadn't affected it at all, he said.

In the pub, this time, I asked if anything had changed. Mr Emmott said it straight: 'I am happy to follow any of the Green ideas as long as they don't inconvenience me.' His sincere and direct delivery bowled me over, but even the strongest-willed activist could see the point. There was a case for convenience, even if change were to be all-pervasive. I told him that ardent recyclers often grumbled at the distance to the bottle bank. Yes, they had cars and designer clothes, were tempted by take-away curry and collected art. And why not? You did not have to be purer than pure to retain a radical, questioning spirit. Mr Emmott thought on the same lines as many activists. Their values were probably similar, as regards thrift and family security and a community future. They would agree that convenience was a way forward.

So what, then, were they arguing about? They were at loggerheads, firstly, because Green people with their sensible ideas and positive solutions were as invisible as Martians to the naked eye, even in Hebden Bridge and Todmorden. Perhaps they were too busy to show a public face. Secondly, if you accepted the need to become Green, you had to accept inconvenience. The Greens would say everyone had to share the effort—and make it individually—to reach convenience. Mr Emmott could, for one, use his business buying power to purchase ecologically-sounder raw materials. But his answer suggested he was reluctant to admit the need for change.

Over the home-cooked, home-grown dinner, he passed around the latest country magazines, advertising the latest Emmott-inspired boardroom tables and chairs. The copy read something like this: 'All our furniture is made from man-planted mahogany and every tree felled is replaced.'

CHAPTER ELEVEN

Rod Everett
& Middle Wood Trust

I read *The Times* over breakfast. ICI announced that they were to sell off their UK fertiliser business. Was that a victory for the organic movement? They had decided to keep the worldwide business, and the British operation would continue under new management. A rearrangement of the deck chairs on the Titanic?

I followed directions from Lancaster station to Middle Wood. Left before Hornby, up the hill, left for a mile, over the cattle grid. There were billowing hedgerows on either side. As I rode on, the bland valley greenery changed into a rainbow of wild flowers. Rod Everett had told me it was a mile up to the cattle grid.

Eventually, at the gate, I found a middle-aged couple juggling with a pole lathe in a hay barn. Their teacher, a Green lifestyler from beard to walking boots, leaned over them to angle a chisel or explain their foot timing on the rope sprung on a long stick. This was Middle Wood Trust in action, teaching an appreciation of natural ways.

Down the track, a stack of grey hair bolted across the field. It was Dr Rod Everett, the trust's co-ordinator. I skipped past some shabby sheds to catch up—a lived-in caravan, a portakabin, a lean-to carport filled with junk. Rod suggested that I go into the house— his newly-built, ecological house, with conservatory—and make a cup of tea.

Inside all was industry. Rod's wife Jane was cooking at the stove. Two volunteer MBA students from Lancaster Business School sat at the kitchen table and earnestly discussed the results of a questionnaire they had conducted to gauge what Lancastrians wanted out of a local environmental centre. A lanky French student sat in the corner. He had volunteered to spend the summer at Middle Wood to learn English, after he had read about it in an international jobs directory. He had arrived two days earlier and looked decidedly homesick.

Rod returned and we moved through to the sitting room to chat. He was pleased to hear that ICI had sold its fertiliser business, though no fertilisers had ever been used on his smallholding near Lancaster. Every bit of his land, from ancient native woodland to rough heather sheep farm, was organic. I imagined it was something like Bruce Marshall's would be in twenty years' time, except for the added experimental vegetable garden and the ecologically-conceived house.

Middle Wood had been Rod's parents' holiday home, an investment and retirement nest. Away at school, away at university, it was the place he always came back to. In 1980, when he was twenty-nine, the tenant sheep-farmer died. He had just finished working on nature reserve management in Cumbria. The family land offered him the chance to put his ideas into practice, funded by the sixty-acre sheep farm and a camping barn for hikers in the woods.

Rod sold a cottage in the village for a handsome price because it was in a prime commuter belt. Around Lancaster it had become the professionals' prerogative to live out of the town where they worked and the land labourers therefore had to commute in the opposite direction for their jobs. So Rod made enough money to build his own house. He lived in a caravan until his finely-tuned plans came together. He had moved in with his wife Jane, a Lancaster graduate, six months before I called.

The passive solar-heating concept took full advantage of the greenhouse effect. A south-facing conservatory warmed that side of the house with direct sunlight on to a dense, ten-tonne wall. The wall then acted as a heat store. The heat could be topped up by other means: a two-kilowatt windmill provided electricity for the house and some cooking; a back-up generator, in the cellar, worked off a small diesel engine and gave off quite a lot of heat. It also produced the electricity for the house, thus acting as a combined heat and power unit. Some canny design work of vents and pipes drew the warmth off around the rooms.

The house was built in local stone from ruins on the farm, which saved on transport and manufacture. Insulation blocks and plasterboard had been used—they have a low-energy cost because they are a by-product of coal-fired power stations—recycled resources.

Wood came direct from a recycled timber merchant, who had heaps of once-used beams for sale and reuse. This made me smile. I had often thought about waste wood left in skips. Presumably it was mostly burnt along with the other rubbish. Yet here was a man who reclaimed it, in the timbers of demolished buildings, or wherever, and it must have been worth his while.

The whole detached house had at least three bedrooms, although I did not have the chance to count. And the total cost, including local skilled labour, was £40,000.

We ate a frugal lunch, finished off with tiny Swiss chocolates. In the permaculture jargon Middle Wood was asset rich, cash poor. In return for an open door to ecological experiments in tune with the Middle Wood ideals, Rod had set himself up as a guinea pig. Middle Wood Green College was up and running all sorts of courses (though the ethical investment weekend had been cancelled through lack of interest—a statement about the North of England?). Middle Wood Trust controlled the land. Some 150 Friends of Middle Wood brought in annual subscriptions of £2,000 a year from as far afield as Japan and the USA. Rod and Jane's income came from the sheep farm and a camping barn where hikers could stop over.

The plot extended to 230 acres in all, most of which was a Grade 1 SSSI conservation site. The whole site was managed for craft products and conservation purposes. Rod told me that their long-term aim was to build up resources for the main Middle Wood Study Centre, which was to be housed in three buildings, each of which would demonstrate different forms of low-energy construction.

Outside I had another taste of organics. Bruce Marshall's land did not have enough garden for a single meal, let alone self-sufficiency, and it looked as though Middle Wood was the same. It was a bleak day and a shower sprinkled down, and I doubted whether the bare hillside was best for a demonstration garden. The herb slope behind the house was scrub, with more weed than prospective culinary ingredients. A permaculture herb spiral, designed to give a compact variety of light and shade plots, had been made. The year before, all the beds had been moved to lay out the big plan, after early trials. The potato field had been left fallow because their precious time went into the task of establishing a varied set of beds. There was evidence of the classic organic gardeners' green manures, buckwheat and lupin. There was companion planting, with onions alongside carrots to deter carrot fly. There were no-dig mounds and carpets to supress weeds.

'We don't want self-sufficiency at the moment,' Rod elaborated. 'We are here as a resource for people to come and carry out their own ideas. Things that go wrong for you teach you as much as those that go right.' In that spirit, the local school had taken up some beds.

Rod sighed to think that, although ICI had sold out, the Agricultural Chemists' Association had just spent £500,000 on a display at Alton Towers theme park to show the positive effects of fertilisers. He sighed also at the slow change in establishment circles, but knew from his strong grip on the pulse how the very higher echelons were aware and clear-sighted. 'There are some very astute operators in the upper ranks, who do know where these issues will take us and are willing to try to move their organisations some way towards that,' he said.

The power-saving lobby was an example, as was the ICI development of biological pest control—insects to fight pests, instead of chemicals. 'A few years ago you would have lost your job for suggesting things that are on the cards now.'

CHAPTER TWELVE

Stephen Blakeway
& Intermediate Technology

Stephen Blakeway was off to Kenya in a week. He had given his Citroën Dyane to his parents and bought ten films for his camera. At the last minute, he was ploughing through the manual for his flip-top Toshiba portable word-processor, part of the package given him by Intermediate Technology, to learn how to send reports back over the telephone. Stephen, a vet for ten years, was going to train a network of African farmers in basic animal health and first aid. 'I could work there all my life in one particular tiny corner of Kenya and achieve nothing in comparison to the world's infinite problems,' Stephen said. 'I could be a stockbroker, or something, and sit there and say: "What's the point, the crisis is so bad we're all going to die." But even if it's all going to end, I can feel I've done my bit. This project should provide a needed service.'

Stephen spoke with a bright optimism. He wanted to retain his idealism but did not want to be seen as a pioneer out to save the world. He laughed at the suggestion, smiled and rocked back on his chair in Intermediate Technology's office. In a few years' time, he would come back and live in a Shropshire village and put down roots in the community—an opportunity missed in a childhood spent moving around with his army father. He would have a small practice and take up invitations to talk in schools and to Women's

Institutes. His rather unconventional slide show might broaden a few horizons, possibly change a few attitudes to the developing world. And he would probably not be asked back. I doubted it. The man had charm and lean good looks, neatly dressed in a striped shirt and cream flannels, tall but squeezed like a caged animal into the cubicle partitions that made up his office.

This was the third floor of the Intermediate Technology Development Group's (ITDG) concrete headquarters, a block inconspicuously sandwiched by Rugby railway station on one side and the town mart on the other. The building was not beautiful, but was cheap office space—about a tenth of the rents in London.Up the road was a playground and shops selling bathroom tiles. Outside the building were an inordinate number of parked cars and just a few bikes. I had entered, hoping that the rows of commuter bikes at the station, and not the cars, belonged to the 100 ITDG workers who promote appropriate technology for the developing world. Did their founder Fritz Schumacher cycle to work twenty-five years before? I wondered.

This was the first time on my travels I had been asked to wait in Reception while a switchboard operator called up to check my appointment. As I waited for Emma Bland, the information officer, a man with a damaged swivel chair told the secretary it was beyond repair, ready for the skip. Surely some valuable Heath Robinson contraption could be devised from the parts?

I tried to be realistic. The visitors' book on the coffee table certainly suggested that I ought to be more respectful. Charles Hoult, Green Books was a precocious nobody compared to... Minister of Industry, Philippines; Minister of Co-operatives, Fiji; Joint Director, Ministry of Heavy Industry, Government of India; Director, National Swedish Board of Technology Development.

On the book rack was a selection of IT books written to impart experience and stop the reinvention of the wheel, which Pat Fleming had seen as so urgent a service. The range was amazing. The sixty-page catalogue covered everything:

The Barefoot Book — economically appropriate services for the rural poor. A key aim is to help people to help themselves, so that they can fully contribute to the social and economic development of their communities.
Bukina Faso — *le secteur informal de Ouagadougou*
Cambio Tecnologico en poblaciones rurales andinas.
Crocodiles as a Resource for the Tropics — *profit while protecting the wild population.*
Goat Health Handbook
Lost Crops of the Incas — other than the potato, the native food

crops of the Andes are largely unknown. Forgotten crops that show promise not only for the Andes, but for other regions.
Small Hydro Power in China.

My call came and I was briefly interrogated by Ms Bland. I explained that I was cycling around the country. I had visited people, places and projects loosely involved in the Green movement: bakers, organic farmers, communes, smallholders. I understood that Intermediate Technology had a British base but worked internationally. I hoped that she could explain the need for this foreign intervention, and tell me who designated where the taskforce went. 'That depends,' she said.

She gave me the leaflets I ought to have read in advance so that I didn't ask simple questions. In 1991 ITDG would celebrate their twenty-fifth anniversary. Fritz Schumacher had challenged the wisdom of million-dollar prestige Third World development, by campaigns to promote the idea of appropriate and intermediate technology. He talked of small scale. By the 1970s ITDG had initiated some projects. One idea was micro hydro: tiny water-driven power stations run by village people at their nearest stream. Villagers owned the power and they could develop at an evolutionary pace. There was no enormous scheme to flood the valley and drive them to the cities. Nor was there the revolution which resulted when a banker paid compensation for the land and left the landless in an unfamiliar cash economy.

The ITDG projects developed when aid workers identified a gap in traditional life caused by outside influences. So Intermediate Technology was increasingly called on to work in the field, to implement practical solutions. Thus in one project, ITDG workers went to help in Kenya's Turkanaland on the border with Ethiopia. They talked to locals—after all, the locals knew their problems best and probably had an idea how they might be solved. Traditionally Turkana tribespeople roamed as pastoralists from fertile valley to fertile plain but lengthy droughts had forced them to settle and wait for rain. The trained agriculturalists from Kenya and Britain came up with an irrigation scheme to 'harvest' the rain. They levelled the land and dammed it with stones so that when the rain came they could use it most effectively, not see it trickle or gush off the caked soil.

'When traditional life is no longer the most appropriate to changes from outside, we try to provide or suggest solutions to what in effect are new and unsolved problems,' Emma told me:

'We give people a choice. They can compare the high-tech, high dollar approach with our evolving, developing projects in partnership with other organisations like Oxfam. We feel if we involve the recipients the help will last and become self-help and therefore sustainable.'

In 1989 the British government contributed half of ITDG's £4.5 million budget. So did the government recognise how useful their work was? 'The government, through the Overseas Development Agency, have many millions to spend, but we have built a wide reputation on this small stake,' Emma said. 'Our work was highly commended in the House of Lords debate in May on development funding.'

For the first ten years ITDG tried to tell governments that small was beautiful. The second ten years were spent concentrating on small as possible. The nineties would be spent persuading governments to run their own small shows.

'If public opinion comes round to development the way we believe in, the government could be changed to our way of thinking,' said Emma, as a parting thought.

She deposited me in Reception again and went in search of a project worker I might talk to. Most of them were overseas for the summer, or were just back or just off and too busy to talk to me, she said. That was when I met Stephen Blakeway, shuffling his papers, eager to pack up and get out of the office. On his desk was a new Toshiba flip-top I could have done with any day he wanted to swap it for an intermediate technology pencil! Beside it was a half-eaten sponge cake cooked by Stephen's mum. It was meant for his first days away but he offered me a piece.

The question still nagged me: 'Why send aid in the first place?' I had thought the argument had been won for partnership, intermediate simplicity and eventual self-help, as opposed to the big boys stepping in with hob-nailed boots. After all, I had sent Blue Peter my milk bottle-tops ages ago for their Christmas appeal to buy hoes and shovels to replant Africa. Hadn't the crops grown yet? Stephen laughed and rocked back on his chair again. His own experience demonstrated how infinitely more complicated life was out there. He was only about ten years older than me, in his mid-thirties, like most people in the office. 'There's a place for this Band Aid fundraising,' he sympathised, 'but once you have the awareness you realise you could do better.'

He explained in simple steps. After a crisis, people need food aid, and quickly. But they could become totally dependent on hand-outs so gradually a 'food for work' programme would start, with the people supporting themselves by exchange of labour for sustenance. Ultimately they would have to break back into independence and wrestle with the choices of an appropriate and viable solution to their new, imposed predicament. 'Even if they are dying, people see that what's more important is preserving their way of life and their community.'

So why muscle in at all, even if you're not saying outright that you know better? 'All these things are very questionable, yes,' Stephen openly admitted. He was Kenya-bound, on one level, simply because he had followed a job advert in a vets' magazine which offered somewhere near the going British rate. And the Schumacher ideas fitted his vision. 'I applied out of a feeling that I knew what was right and what was wrong, and this was right,' he said. He had enjoyed travel and seeing the world as a student at Cambridge University, too, though after a while he had realised that one learns more if one works there rather than just passing through.

After university Stephen had worked in Herefordshire, in a country practice, to get a grounding. But he soon took a job abroad in Papua New Guinea. In three years as a government area officer he had supervised animal disease surveillance and trained government para-vets. There were no more than half a dozen vets in the whole of Papua New Guinea, and a couple of them just treated expatriate Australians' pets. Stephen and two others monitored the rest of the country's livestock. A missionary in the Highlands did something similar on his patch. Stephen's main job was to train government agricultural extension officers who went out into the field to act as para-vets with skills to treat a few common diseases. As a result the country had been released from the grip of common epidemics.

Unfortunately the civil servants had a Western model for their work and a colonial heritage. A government trip required the pomp of an official car. They demanded travel expenses, overnight allowances, mileage, petrol money. Simon found their mentality entirely inappropriate and at odds with the state of a country where 85 per cent of its people lived in villages. Everywhere people were short of cash and there were only a very few conventionally conceived 'jobs' for the élite, who had been alienated by education.

He had returned to England for a stint so that he did not lose touch with British ways, but he hankered after another challenge abroad - hence the job in Kenya. 'Extension' was also the key to ITDG's Kenya project. One could now take courses in the subject at Reading University or Wye College, Stephen said. It was new jargon, and a new concept to me. The idea is that if there are not enough qualified vets, then those vets should share their knowledge and teach basic skills to a few members in each community. The new specialist lay people could then pool their combined skill to set up a local network, perhaps with a drug store. The ITDG plan was to train para-vets from each area so that they had no hang-ups about travel expenses or petrol allowances—they were helping their friends, who would in turn help them.

Why aid? Why Kenya? The conflicts nagged. Was this just a good jaunt for a young vet? No. The chain from food aid back to subsistence and self-sufficiency had been established. The Kenyan government had a good hold on the big problems of epidemics like foot and mouth disease, but the small-time ailments were not being reached. Animal health ideas were catching up with human health ideas. In Britain, most vet work is now at the level of individual work for one-off problems—James Heriott stuff, not whole herds with a plague. Stephen would have only two staff, a Kenyan and a European, and a tiny budget from ITDG. They would run six centres, alongside Oxfam projects which were running women's groups and irrigation workshops. Oxfam had called them in because they saw the gap in animal health care. Kenya had been chosen because things were going on which they could plug into and because far from feeling threatened by such projects, the government had decided that the idea was excellent and they would monitor it closely.

No money would be splashed around, either, for thirsty men in high places. The trainers would get wages, but the Kenyans could set up shop, sell drugs and cure animals as soon as they had the basic ideas. The only expense might be the cost of a few trips for learners to visit the more advanced set-ups. The idea was already well grounded in the skills on offer and the needs which had to be addressed. I was convinced that the project could only do more good than harm. It was in no way pious and it would leave behind a potentially sustainable and simple system for animal health.

CHAPTER THIRTEEN

The Henry Doubleday Research Association

I t is said that American lawns receive more artificial fertiliser than the whole of Africa's crops. Like Rod Everett's Middle Wood Trust, Ryton Gardens, just outside Coventry, used none. What would you expect at the Henry Doubleday Research Association, alias the National Centre for Organic Gardening, the Kew Gardens of the Green movement?

I had rung Sue Stickland, the head gardener, several times to arrange a visit by myself and John Homer, secretary of the local FoE group. Each time she was out in the grounds. So John and I had left a final message and set off for the short ride to the outskirts. To have someone to chat to on the move was a treat. To be led, rather than to scramble with my map, was another.

John knew the way because he had joined the Henry Doubleday Research Association, an organic gardeners' question and answer service, when it had expanded and moved North. The gardens were just fields when he had last visited four years before, so he was interested to see how much had grown.

The founder, Lawrence Hills, had sold the Association's two acres of land at Bocking, Essex, for development in 1985 and bought a twenty-two acre former riding school with the money. The land had not been treated with artificial fertiliser for a long time, so the expanded HDRA team had not had to cope with chemical residues. They took out a massive bank loan and opened to the public. Lawrence, the man who put comfrey plants on the map as a gardener's panacea, had moved to a cottage on the site, but had handed day-to-day running over to Alan Gear. Alan had set about enlarging the scope of the centre to persuade more gardeners that they would lose nothing if they cut back on chemicals. He had also started a research station to find out the exact empirical effects of organic techniques. Scientific precision was insisted on so that the results would have maximum credibility. In a corner, out of the public glare, stood acres of green manure, carrots and cabbages, and row upon row of different vegetable varieties, studied so they perform under organic conditions.

Expansion was not just in extra space. Membership of the HDRA had more than doubled from 7,000 in 1985, even at £12 a year. Alan Gear became quite a gardening celebrity through his natural knack on radio and television; Channel Four came along and filmed three *All Muck and Magic?* TV series with the growing season telescoped down to eight weekly shows.

The first thing I noticed about the place was its similarity to a large garden centre. The image was a professional one, more like Intermediate Technology's than Rod Everett's. The reception hut was stocked full of seeds, plants and gardening books, organic wine and vegetables. Café staff wore uniforms. Honest respectability on a par with the National Trust was everywhere. There was nothing raggle-taggle about the gardens, though they were somewhat dwarfed by the shelterless expanse in which they grew. Herbaceous borders in large swoops curled around the vegetable patches. Formal gardens and fruit trellises: each had its own plot with yards of parched grass between.

I joined the tour, to have everything formally explained. Eric Barnes, a cheery Falstaff character, led the assembled company through the basics, but as we went along it was obvious that several handy and green-fingered couples would keep him on his toes with astute and intricate questions. We started with composts. Leaf-mould soaked in urine would rot down more quickly. Worm boxes for household waste should be filled with the brandling not the common earthworm. Everyone chipped in with inside knowledge of leaf-mould. The whole entourage had come for high level talks on state-of-the-art gardening. Another hint for household organic waste: you can dig a trench for winter or late beans, throw in the eggshells, peels and teabags, then plant beans over the top because they don't worry about the strong rot mix around the roots.

We saw alternatives to peat. 'Don't use peat, it destroys the environment,' said Eric Barnes. 'Before, when we had enough, it was OK. Now, with great digger machines, they can take up the whole habitat, down to the rock.' Coir (coconut fibre) did the same trick as peat.

The next beds grew onions in three strips: one end of the soil was completely unnourished, the other well husbanded. Already you could see the poorer yields. 'Don't forget that in a single teaspoon of organic soil there are more living organisms than humans on earth,' Eric said to us and theatrically brandished a handful of dark, urine-soaked sludge he had prepared earlier to feed to the onions.

Eric's confirmed laziness, mixed with a desire not to disturb even a teaspoonful, meant he favoured a no-dig garden, and that was our next stop. The beds had never been dug, but were producing healthy-looking vegetables. Eric shepherded us across the open grounds. The discerning eyes of the group appreciated the formal beauty in patterns and symetries. I wondered how Rod Everett's higgledy-piggledy permaculture would go down with them. Challenge or lost cause? The Rose Garden celebrated form not function; the Bee Garden had a planting theme to suckle the hive that buzzed behind clear perspex; HDRA members' experiments on a tiny corner oddly blended in. The soft fruit cage promised good scrumping for raspberries in the knowledge no one had sprayed anything deadly on them.

One fifteen-foot square displayed a vivid shock of blue flowers over interesting jagged silver-lined leaves. This, Eric explained, was phacelia planted as a 'green manure'. Instead of using complicated leaf-moulds and fiddly, nettle-juice fertilisers, the gardener had planted a crop chosen for its capacity for improving the soil. As the plant grew, weeds would be squeezed out and a neat cover restored for insects and animal life to keep down pests. There was strength in diversity. Clover, of course, was the traditional green manure but phacelia had apparently given the avid gardeners new colourful inspiration because they had emptied the shop of every last seed. I had to make do with crimson clover and winter bean in my expensive spree later.

We stopped at The Ryton Garden, planned as the quarter-acre behind a suburban house. It was a diverse combination of all the concentrated, individual efforts seen in action so far. Flower beds surrounded a large lawn 'leisure' area. Espalier-trained fruit trees shielded off a vegetable patch big enough to feed a family. The vegetable patch was divided into four squares, each of four beds. Crop rotation meant you moved cabbage, potatoes, onions and carrots around year by year.

The gardeners could not have stage-managed the insects buzzing over the cabbages. The haze and variety instantly dispelled the rather sterile impression I had I had so far. What impressed me most of all was how nature had responded to the vibrant man-made diversity. There were no other gardens for miles around and the insects had homed in on this one, making it a cacophony of life. Four different species of butterfly danced in the air. Bees and wasps passed on their way. A horsefly dangled on a stem. A bright ladybird shaded behind the leaf. Finches and sparrows, as precise as Red Arrow jets, fizzed between all obstacles to scoop up a froth of gnats and midges. A truly natural balance had grown as new competitors expanded the ecology.

My tour, by now a conference, fell to tales of pests. The tourists took turns to extemporise with glee on such subjects as club root and carrot fly. It was quite beyond me but each received a chorus of agreement, followed by 'A similar thing happened to me.'

I marvelled at the peacock butterflies brooding on teasel. We were at the Wildlife Garden now, with brambles, wild flowers on low-fertility soil and a hedgehog box (hedgehogs love insects). The pond had a special sloped edge so that hedgehogs did not drown and could crawl out after a dip. There were no fish as yet.

Eric had joined the team at Ryton after being made redundant. He had helped to plant round the lake in the conservation area and was waiting for fish to arrive like the insects. He was prepared to give it seven years. 'They just fly in,' a Dutch gardener said to everyone's surprise. He had, by then, scored many points on our tour after a quiet start. 'Yes, they are flown in by duck.' The theory was that fish eggs were caught in the ducks' webbed feet and brought in when the ducks came to feed on the plants. Eventually some eggs would survive.

Eric, bearded and bucolic, clad in a checked shirt and stocky walking boots, discreetly took tips and went back to the HDRA caravan. My messages to Sue Stickland had got through and so I took a few moments of her busy outdoor day. Compared to Eric's *acteur manqué* performance, she took a more intense line. She was petite but her hands bulged at the joints from the cold and the constant clawing. She pondered my enthusiastic manner.

Sue was a latecomer to gardening. She had published textbooks for the Open University Technology Faculty after taking degrees in physics. Friends would ask her to help if they wanted to publish something. She had got involved with *In the Making*, a directory of co-operative businesses, and she wrote for *Undercurrents*. 'You've heard of *Undercurrents*?' she asked me, impassively.

'Of course I have,' I replied. Thank goodness I had met somebody at Glastonbury who knew somebody who knew somebody else who had been involved. It had achieved cult status since *Resurgence* magazine had taken it over.

Sue had eventually decided to abandon the pressure of the publishing world and had got a job in the field with Milton Keynes Development Corporation. She had shown schoolchildren the nature around them which would survive when they covered the landscape with new town roads. She had then joined a task-force harvesting the wild seeds and plants before bulldozers hacked up their soil.

Having made a break from the office she had felt that she couldn't return, so she had taken a horticulture course at Lackham College, Wiltshire—and had been hooked. It was outside, creative, different every year. She had tended a large merchant banker's garden in Tetbury, Gloucestershire, then moved to head the HDRA hierarchy when operations moved to Ryton, to organise the gardens with three full-time gardeners tending displays for 33,000 visitors a year. She unassumingly graduated to one of the TV presenters for *All Muck and Magic?*.

A stroll away, we were in grids of cabbages, leeks and carrots on the research fields. Three gaunt, bronzed bodies bent over a cabbage patch all in crisp straw hats. Lawrence Hills had set a fashion for the hats. A German student, clipboard in hand, noted the weights of the browning cabbages under different green manure crops. The chief scientist weighed them. A summer volunteer from agricultural college sliced them off at the stem with a knife and threw them into piles.

The increase in research has been exponential as organic gardening has taken off. The HDRA has outposts in Spain and the Cape Verde Islands. Among other things they have experimented on strains of trees to defeat droughts. And the grants have flowed in. The National Agricultural Botany Institute has commissioned experiments, as have the Ministry of Agriculture and seed companies. This was the stuff Dennis at the Watermill in Little Salkeld had grumbled about. You planted beans a caliper-measured three inches apart and never stopped gauging.

The soft-sell had succeeded better than any garden centre with rows of wilted varieties. The shop had drawn me like a museum shop because something called out to say you wanted to own what you had seen. Consumption was in the blood. I bought seeds with no idea where I would plant them, and a bottle of organic wine as a present for John Homer.

CHAPTER FOURTEEN

Friends of the Earth, Birmingham

The newspapers screamed about a 99° heatwave, and shuddered at Iraq's manoeuvres in Kuwait. Petrol prices would rocket. I pedalled furiously and arrived in Digbeth at Birmingham FoE's warehouse. I parked the KP Crisps machine on the indoor bicycle racks bounded by stacks of recycled paper sheaves. Big sheets for printers, rainbow pads for schools.

Paul Stephenson, Birmingham FoE's paid information co-ordinator, rushed past and promised to return after doing a spot of business. Paul had spent a year with a finger in every FoE pie from collecting paper for recycling to the Tropical Rainforest Campaign. He knew rainforest campaigners back from photographic assignments to the Penan tribe—he probably even knew Jonathon. He introduced Graham Lennard, just back from a month's cycle tour of the Deep South of America with his wife. Ninety-nine degrees? Pah. And humidity...? Nevertheless I had fried on the triple carriageway between Coventry and Birmingham—and I guzzled cups of water as Paul explained the set-up.

In the 1970s West Midlands County Council had pumped money into Birmingham FoE through the Community Programme to give work to the unemployed. The warehouse had hummed with a bike workshop, recycling and a large company of home-insulation fitters, and FoE had had the man-management headache of workers, committed or otherwise. In the break-up of the metropolitan county, some left-over cash had come FoE's way. They had bought the building, which had additional space for sympathetic businesses. Paul showed me through One Earth wholefood shop and the offices of CND, the Green Party and the National Association of Widows. Tucked in the attic, Paul shared his office with Recycled Paper Supplies, a co-op run full-time by Graham's wife Liz. Pushbikes and various volunteer wings behind FoE campaigns shared the building. Dene Stevens had dropped in to sort through orders for stickers and badges. He handled national badge sales to FoE groups, which earned extra money for the autonomous Birmingham branch.

Everyone was wilting. The zealous insulation teams of yore had lagged the roof so well that there was no escape of heat on a hot summer's day and no draught through the open fire escape. As I'd come through the back maze of paper I had missed reception: it was not the bland soulless sofas of Intermediate Technology, but a cosy pine-panelled treasure trove. Chris Brimson, the morning's volunteer, cradled the phone whilst surrounded by briefing packs and other useful material, amid precarious pot plants and a clutter of books. Racks of leaflets, each with a nominal charge, were stacked high for access at a glance. Behind the desk were shelves stocked with videos from Ark or rainforest groups. Photo boards leaned against the walls. Information was the frontline effort.

Paul and FoE regularly received requests to stage a show. Even 'respectable' people, like Bovis Homes, the Midland Landscape Architects' Association and the Institute of Chemical Engineers, now wanted to hear the FoE line. After the talk they would ask Paul when he would look for a proper job and thank him for his interesting perspective.

The phone rang, and Chris handed an enquiry from Leicester over to Paul. A lady wanted to know how best to complain about her neighbourhood chemical factory. FoE Birmingham, as a self-governed local group was only supposed to deal with Birmingham, Wolverhampton and the Black Country. They only had 600 members in the area affiliated to them with their £5 subscription. Over 280 other groups criss-crossed the rest of Britain under licence to deal with the other 180,000 members of FoE UK, but somehow Paul took calls from everywhere anyway. He had a spate of calls from the Channel Islands and callers from Sussex were not unusual.

'They want us to be a panacea to everything. If they can't think of anyone else, they call us. We're like International Rescue. "My tyres are melting in this heat, what can you do about it," one asked. What did she want me to do? Go down to East Grinstead and sort them out? 'I get asked about rats in the kitchen, slurry in the pond, everything.'

Paul proudly showed me through to the brand new meeting room in a £30,000 extension behind the old warehouse. A graffiti mural, signed 'Ozone Friendly', lined the ten by fifteen foot wall. It came from a concert in aid of FoE which some fashionable crowd had organised independently. We tripped over a gang of children on our way back. They clutched passports of the Birmingham Action Adventure game, a summer holiday activity run by the council with sixty stations across the city to visit, and with prizes for the most stamps. At FoE, the children had to colour in a piece of cardboard jigsaw 'environmentally'. I was amazed by the initiative of Imogen and Lara Featherstone, the two young sisters who had dreamed up the stunt and liaised with the council to register their stop, set up a rota to take charge and invented an environmental quiz for participants. The giant jigsaw would be displayed at the November prize party. 'Then recycled, of course,' Imogen and Lara chorused.

The sisters had joined FoE's junior section, Earth Action, the previous Christmas. 'Imogen wanted to join a Green movement.' Theirs was the seventieth Earth Action group and was started by Nancy Hine who got things going for the younger crowd. Nancy was a final year university student at the time but had passed the twenty-three-year-old age limit. Sheepishly, I admitted that I hadn't—but only to Imogen and Lara.

'Nancy organised a national Earth Action conference on how to do action days and help demos,' Imogen told me. On action days Earth Action gangs joined the older FoE campaigners to protest against a specific issue, be it the greenhouse effect or pollution. 'It was the best thing we'd done. We learnt how to phone the press, etc. Only forty came because FoE had to subsidise the train fares, but there was a disco at the end.'

Imogen was dauntingly precocious for a fourteen-year-old. What use had I been at that age? She and Lara came from Kings Heath each Saturday to meet the others from as far away as Sutton Coldfield to plot action. 'We get costume ideas from the newsletter, but our own are much better!' Lara trumpeted with excitement. They've dressed as loo rolls and trees for past demonstrations. On the last FoE action day, a national ozone layer event, they dressed as twenty-first century sunbathers in white boiler suits, head gear and sunglasses with sun block on the exposed areas of flesh.

Monday evening was Birmingham FoE's big night, Paul told me over lunch at the Amazon Café. Each campaign committee met for business and feedback. Co-ordinators typed their letters, small huddles schemed about stratagems for conscience-raising. The committed rose through these ranks. Some were on the dole and between jobs so they could give time. Others had demanding jobs so they gave money. A core of twenty regularly turned up. Chris Crean, the chair, was reputedly elected on his first visit.

Paul had come back to Birmingham after taking an environmental science degree, having had some nature conservation experience in Hertfordshire. He had stuck with FoE when he was unemployed and earned his share of any grants that came their way. Now he was the lucky one with a one-year full-time job paid for by FoE headquarters. Others not so lucky still put body and soul into the cause and hoped something similar would come their way. It had for Barbara Norden, who was on a local authority contract to promote cycling in the West Midlands. And Graham Lennard had scraped together a year's salary in 1989 to edit *A Green Guide to Birmingham*. I had not realised he was responsible for it until then. His book had been an inspiration to me for ages, when I had thought about compiling a similar work for my home town. I did some mild hero-worship of the man when we cycled back to his house with Liz after work. There was nothing more satisfying than to ride in convoy with fellow cyclists on a sunny Friday evening— especially as these ones had offered to put me up for the night.

Graham's *Green Guide* had come out around the time of the *Green Consumer Guide* and dealt with similar issues in a comprehensive, city-wide directory. Unfortunately the much-hyped *Consumer Guide* had rather swamped the Birmingham one and sales had been sluggish.

I heard about the Green comings and goings in Birmingham over a Hindu vegetarian meal at Jaaneman's restaurant in Sparkhill. Radical Roots and the New University sounded like the early Red Herring in Newcastle—a similarly enigmatic bunch of militant ideologues and a few others committed to actual graft. Ashram Acres, an inner city housing co-op with control over various allotments and back gardens, grew Asian and Afro-Caribbean vegetables and kept animals for milk, eggs, furs, honey.

Graham and Liz told their American cycling tales of Deep South hospitality and lifts they had hitched when the temperature was too high, with the bikes thrust in the back of pick-up trucks. I wished some empty lorries had shown me such kindliness. Graham added an interlude to explain why he had a bicycle in the bathroom. It was a Moulton frame, one of the small-wheeled variety which had no street credibility when I was a child. Graham had rescued his mother-in-law's because the old design had taken on some collectable, connoisseur value and stowed it in the bathroom awaiting repair. He had ordered new wheels from the Midlands secretary of the Moulton Bicycle Club, in Wolverhampton. The club was in the grand tradition of owner eccentrics, with tales abounding of freak performing bikes and near-archaeological finds of pristine, vintage machines. It was dedicated to the worship of Dr Alex Moulton of Bradford-on-Avon and his brainchildren, the latest of which, with unique nine-cog rear gear block (selling price $4200), had just come eighth in the killer cross-America race.

Chris Crean arrived late with an evening swimmer's appetite and ate our puris. We talked about water because his business was its cleanliness. He was involved as a consultant with the water privatisation and with the single European market in 1992, and advised clients with their own springs whether to keep them going so that they could offer competition after privatisation. At the same time, modernisation was required to meet the new European Community (EC) standards.

'Of course, it's ridiculous that we use most of our water to flush our toilets, and we use drinking quality for it,' he said. 'Why on earth don't we have plumbing to re-use water that's hardly dirty from the sink or bath?' (I later discovered that we also lose half that purified water straight back to earth through cracked pipes.)

I mentioned the permaculture idea of rainwater collection. It would become increasingly feasible in Britain as EC standards nudged up prices. 'But,' Chris pondered, 'it might be too pure to drink straight. The body needs water with minerals in it otherwise the water starts to draw them out.'

Alcohol-free Jaaneman's shut at eight o'clock, and we walked to the Malted Shovel for a pint. With a round of Burton Ales in the garden, the conversation loosened up. The young crowd of friends swelled and the ring of plastic seats filled with teachers, research scientists and a couple of jugglers.

Chris expounded on the price of petrol (his mountain bike safely locked to a drain pipe in the beer garden). Iraq's annexation of Kuwait had increased it, but the market jitters would be good for alternatives to cars. The same applied to the Hanson bid to buy the PowerGen business from the government, instead of shares being issued to the public. 'Of course, we're pissed off,' he said, 'because we know he's going to get a bargain. But because of the need to satisfy shareholders, the costs will make consumers think.' In America, for example, private electricity businesses actually found it made economic sense to promote energy-saving devices so that they did not have to build more power stations. It was better to share and sustain the output than be forced to invest in half-capacity generators guaranteed never to make an effective profit.

A combined heat and power (CHP) generator would become the cost-effective option. Birmingham's new Greenfield Hospital had commissioned one. I'd seen Rod Everett's own machine at Middle Wood. 'This is much bigger, like the ones in Holland,' Chris said. There, CHP technology supplied whole villages or suburbs. The management company could nearly double energy output for the same resources because, as electricity was wired into the circuit, heat was pumped into the central heating. The consumer got a better bargain but the incentives to save energy remained in the price-to-consumption relationship.

As the night darkened I talked city allotments and open spaces with Liz. She had opened her mail that evening to find that the officious allotment attendant had requested that she remove weeds from her plots. 'Mine are artichokes,' she laughed. Onions and potatoes had been successful crops in the past, but Liz had adopted the principle of growing what's expensive to buy, like courgettes and aubergines, or best eaten freshly cut, like lettuce.

Did money affect their lifestyle when they were paid co-operative wages, then? Liz dropped in some ambivalent thoughts of university friends. 'A few years ago we all had first jobs and we led similarly thrifty lives but now we are going on thirty and some have bigger cars or have moved house,' she said. She and Graham bought their terraced two-up at an affordable £15,000 in 1984 before prices took off. They would have found it impossible to move into the property market later. Equally a step up to a larger house in Mosley was prohibitive. Split-second eye contact confirmed that Graham had overheard.

Graham was a more experienced cyclist than me and he would take a longer, quieter route in preference to a shorter, busier one. He encouraged me to take a train to Wolverhampton instead of battle the Saturday shopping traffic out to Shropshire. A detour took us past the communal houses of Ashram Acres. I glimpsed their 'poly tunnel' hot-houses. I had joked over my muesli at breakfast that the difference between Ryton Gardens and Middle Wood Trust was that Ryton appealed to mainstream horticulturalists whereas Middle Wood was more pitched to confident types who cared less about imposed, conventional order and painted their lintels shocking colours. The Ashram Acres window frames shone orange and yellow but their burly guard dog did not give me time to wonder about them before it hurled itself against the fence, then burrowed below.

Graham joined me on the Wolverhampton train. It was the big

day for him to collect his new Moulton wheels from the Midlands club secretary. The secretary himself had gone to Milton Keynes for a club ride with his brand new bike but his wife answered the chiming doorbell. She, husband and son often took a Moulton each from the family's collection of five and cycled off to Wales for a youth hostel weekend. The eccentricity of the bicycle designs and the strange fervour of these Moulton apostles made for an indescribably comic impression. But I bit my lip and smiled.

My amusement reminded Graham that there was 'naught as queer as folk'. A lady from Richmond had written to him requesting samples from Recycled Paper Supplies. She had called back when I was in the office. 'I've recently gone Green, you see,' she said, 'and I need the paper for my social stationery, but yours is not quite right.'

CHAPTER FIFTEEN

Robert Hart, agroforester

R oman Bank, behind Robert Hart's house, finally defeated me as I climbed the steep shadeless road and the sun burnt down on me at its most intense since Wolverhampton five hours before. A buzzard soared overhead and disappeared. I was quite prepared to believe that I had seen a mirage, after straining up the long hill. Going down after the push, a cool breeze soothed my aches for a few seconds before I reached Robert Hart's forest garden and agroforestry experiments on the other north-west facing side.

I followed the track which pilgrims had taken before me. The *All Muck and Magic?* TV crew had been there for a day, as had Bill 'Permaculture' Mollison and even Jonathon himself. They had come, as I had, to see the culmination of Robert Hart's work on self-sufficiency: an eighth of an acre planted in storeys from tree canopy to root-crop level. The garden had received an award from the Institute of Social Inventions in 1987 (Richard Adams' New Consumer won one in 1989) but, many years earlier, Hart had turned to a mixed agriculture centred around trees, berries and perennials.

I knocked at the house. A man redirected me behind the garage. I squeezed through bushes into a wilderness with tilted fences, fruit trees and a morass of greenery flopped on either side of straw-laid paths. I could not find the door so I clambered over a few hay bails and tapped on the bay window. Robert Hart came out of nowhere and offered me his hand. I shook it. Without a word in answer to my apologies he ushered me indoors with the motion of a seed sower.

The front room was small, the bay window being shadowed by an outgrown pear tree and sprigs of fennel. The cill was book-lined and cobwebs crossed where one edition dwarfed a neighbour. *Arguments for Democracy* by Tony Benn was strung to *Arguments for Socialism* with a smaller book in between. I chose an antique chair with a worn tapestry seat and sat down. From under a freshly laundered tea towel my host unveiled a feast of fruit and salad. I drank apple juice and we made acquaintance, though Mr Hart would occasionally and disconcertingly nip out to tend to his brother, who was unwell. When he was out of the room I looked around the pink-washed walls. Family miniatures and photographs of former homes hung on two, while a third was pasted with letters. 'From Indira Ghandi, Prime Minister, India,' one read. 'I have received your copy of *Ecosociety*, thank you.' It was dated October 1984 and Mr Hart reckoned it to be one of her last. A chicken-wire frame was suspended on ropes overhead as a herb dryer. From the rust and the shrivelled brown leaves, I reckoned it was largely idle now. Beside it hung a wrought iron candelabrum in the shape of a ship. 'It is reputed to have come from a Venetian palace,' Mr Hart told me.

His forbears had strongly shaped his thinking, Mr Hart went on:

'Robert Hart the First was a steel engraver, a skilled craftsman who came from Melrose in the well-wooded Tweed valley. I have one of his finest engravings: a portrait of John Evelyn whose great work *Silva* led to the planting of millions of trees in an England denuded by the "great bravery of building" and ship-building during the Tudor and early Stuart periods. Robert had a passion for history and, in particular, for the story of mankind's age-long struggle for freedom. He named his two sons George Washington and John Hampden and his three daughters after ladies who had all met violent deaths in the cause of freedom: Boadicea, Lucretia and Virginia.'

He seemed to find some symbolic appropriateness in the past. John Evelyn had written about salads and Robert Hart, his mother and his brother had lived off the recipes for several years. When they settled in Shropshire, they discovered that relatives had inhabited the valley centuries ago and this made it feel a better place to live.

I finished the apple juice and was offered local spring water. I realised, when Mr Hart fished out several blades of grass, that the green plastic bottle had been filled up by hand not taken from a supermarket shelf. The contents tasted sufficiently of nettles to rumble my bowels. 'Oh, would you mind using the outside lavatory, though?' Mr Hart answered my request. We wended our way to the bottom of the garden. Behind the summerhouse, complete with a railwayman's cooker, was an outside toilet: a bucket with straw at bottom and another with a brush and rainwater to slop out onto the compost. Mr Hart chopped wood for the stove, his only cooker, and fidgeted as the time ticked away and I did not appear from the commode. Never have curry before an interview!

It was food that began Mr Hart's own odyssey. He developed an interest in the world food crisis when working as a Reuters journalist. And after his father, an international lawyer, died, he left London to write a book on the subject, fleshed out by first-hand experience. He bought a farm in Somerset and moved his ailing mother and brother to live where nature could be their remedy. Steadily, he learnt the country ways and discovered that the plum trees were undersown with blackcurrants in his orchard for a special reason. It was a traditional association, his first introduction to 'plant symbiosis' or companion planting.

The writings of John Evelyn helped him decide on his crops because he was trying to simplify and purify the family diet. As I sat and listened to this articulate introspection, he produced a facsimile 1699 version of Evelyn's *Salets* from the dusty bookshelf. Joy Larkcon, a world authority on salads, had given it to him, he told me.

In 1960 the Harts moved to Wenlock Edge in Shropshire and Robert put his back into farming twenty acres of grasslands with the help of three old-fashioned countrymen brothers who lived in the village. They reared dairy heifers from the herds of Sam Mayall, a Soil Association founder. They planted two orchards and a herb garden in place of the lawn. Everything he learnt came from the three brothers.

'It was a way of life for my brother and myself. The aim was partly self-sufficiency. I wanted us to be as far as possible self-sufficient in fruit, vegetables, herbs, milk and eggs. Selling cattle was the cash component and this farm became known for its cattle. I went on to single suckling calves, where the mother feeds the calf for the full nine months and weans it herself. It produced the very best.'

All the time, Mr Hart wrote about organic farming and the world food crisis. He published *The Inviolable Hills* in 1968, with a preface by Lady Eve Balfour, then secretary of the Soil Association. His ideas were based on early experience of companion planting as well as perennial vegetables and herbs. 'It's really rather topical now,' he said, 'because it is about integrating farming with forestry for the sake of conservation.'

While he was writing this book Lady Eve sent him an article by James Sholto Douglas about Kagawa, a Japanese Christian evangelist who modelled himself on Gandhi. Kagawa had instituted 'three-dimensional farming'—trees for conservation and feed for animals—and Sholto Douglas copied the method on his farm along the Limpopo valley in Southern Africa. His livestock were nurtured on bean-bearing trees such as carobs. Mr Hart told me:

'I wrote to James Sholto Douglas at once and tried to adapt the system to temperate conditions. Lawrence Hills said at the time that it was impossible and that temperate forest farming was a contradiction in terms, but I decided it wasn't. There were examples of traditional agroforestry in this country as all over the world.

'For instance, one component of the Saxon village community was the oak wood. Villagers developed a section of the primeval forest in a sustainable way using the trees - pollarding and coppicing them for timber and stakes and fuel. They used the acorns to feed their swine. In fact, if you read the Doomsday Book you will find that in many villages they mention the oak wood. And the standard for an oak wood was that it would feed so many swine. Here, this village Rushbury had a wood to feed forty swine.

'The Saxons also had an encyclopaedic knowledge of herbs in the forest, so that was another system of agroforestry.'

The experiments continued and Mr Hart became a vegan because he felt it was healthier. So he let off his land to other farmers and his house to a commune of some fifteen people who camped around in caravans. He concentrated on growing health-giving herbs and establishing agroforestry.

'I realised that herbs and perennial vegetables were far less labour-intensive. I realised that if I was to be effectively self-sufficient I would have to devise a diet which eliminated annual vegetables. I called it ecocultivation, and planted symbiotically.'

About this time, he came across Bill Mollison's ideas and also a system evolved in the rainforest of Southern China. Their independent experiments showed that in ten years rampant, near-natural growth showed higher productivity returns without any inputs of fertiliser or water.

In 1976 Sholto Douglas and Hart wrote *Forest Farming* together. I said that I had seen the book still selling well at Glastonbury on the permaculture stall. Mr Hart smiled. It made him about £70 a year after sharing the royalties 40/60 with Sholto Douglas. He was more interested to see how the ideas cropped up all over the world without reference to the book. A Canadian scientist had even founded an International Council for Research in Agroforestry in Nairobi, Kenya, to set about unearthing agroforestry's many forms in many countries.

Mr Hart decided to develop his own garden-sized forest. A good start to solving the ecological crisis would be for people to plant in their own back yards. One tree in each would grow a revolution. With the help of Garnet Jones, his gardener, he worked on a forest garden.

As the temperature outside had cooled we went to have a look at the wilderness. The garden was well established, with some trees over fifteen years old and the high hedge surrounds much more ancient. Late summer growth had gone from lush to seed-laden, so large plants drooped and gave some borders a thin look. Everywhere were apple trees. I remembered: 'The greater the complexity of the whole, the greater is the energy and fertility which is generated.' There was energy everywhere. To Mr Hart, the landscape was much more than his food. He had imbued it with a spirit founded in the minutiae of history and traditions, of archaeology and fable. His hermit life alone with his brother had forced his roots deep into the land. The casual suggestion of a Norman motte and bailey fort spawned a mythology about the lie of the land. The gorge of a stream was perhaps a prehistoric packhorse track, the dip in the field a monastic fishpond.

His early work had kept vegetables, herbs and forest experiments separate, but you could not tell the difference any longer. Nor were there any particular paths. Everywhere was groaning with plant life. We stopped at an edible rowan and a juneberry. 'I quite like the rowan although the berries are bitter,' said Mr Hart as he passed and lent forward to take a close look. 'The North American Indians eat juneberries as a subsistence food.'

A tree beside them had three varieties of apple grafted on to the same trunk. 'It's amazing—when you graft on to an old tree that hardly produces good fruit any longer, the graft will rejuvenate the whole tree.' Still close to the house, below the trees, was an undergrowth of herbs that hated the cramped surrounding of the forest garden, and therefore had their own 'anti-forest'. Mr Hart also called it his 'ante-forest' garden, because his initial experiments were planted there: the sensitive herbs of thyme, yarrow, lemon balm, marjoram and rue.

All over there was catmint and ten other mint varieties which scented the air as we brushed through. The opposite bed was a drenched bog, lined out with plastic sheet so that cranberries and bilberries thrived. Beside them cloudberry grew, together with azerole, an edible hawthorn, grew, giving slight shade to my favourite of vegetables, broccoli. This, Mr Hart explained, was a special type of perennial broccoli. The sprigs can be cut off at any time, and the plant continues to grow. Great claims were made for sea-kale, too, 'the hardiest plant in the world', which grows on the seashore in western Scotland. 'You blanch the leaves for a few minutes before cooking it with something.'

All over the tussocks and twists of the garden sprang unusual plants that bore fruit or berry or leaf ready for the daily salad bowl. Mr Hart had established Good King Henry, a perennial like spinach, which also produces a cereal crop. In the poly-tunnel, 'on the site of the monastic fishpond', he had tried to blossom two kiwi fruit trees, which flourished but bore no fruit.

'Do you have green fingers?' I asked.

'I've got ideas,' he replied. The work was left up to his partner and gardener, whom Mr Hart described as 'a Celt of magnificent physique from an old yeoman family in the wilds of mid-Wales.'

Ideas flourished most in the arboretum. Mr Hart had a collector's mind for the stories behind each species he had planted and delighted in seeing them so close together—whether or not they could possibly survive to maturity in such a confined space. Each tree was dedicated to the memory of someone influential in his life—though in his case none was in memory of women who died violent deaths. The Southern beeches (*Nothogangus antartica* and *procera*), trees that grew in the world's southernmost habitats, were planted in memory of Edward Adrian Wilson, a naturalist and explorer who died with Scott in the Antarctic, and Victor Java, a Chilean folk singer martyred in 1973. The Himalayan whitebeam was chosen to honour Murlidhar Devidas Amte, an Indian 'whose heroic struggles on behalf of leprosy sufferers, of Indian aborigines and their rainforest home has been of truly Himalayan proportions'. My host was a long-time correspondent with Amte and was particularly interested in India, where his agroforestry ideas had been taken seriously, with the publication of his books there. The Japanese red cedar commemorated Toyohiko Kagawa, the 'universal genius' influential in James Sholto Douglas's ideas on forest farming. There were trees from every continent.

Finally we had come round the smallholding to the forest garden itself: an orchard canopy above, shrubs and climbers layered below, root-crops burrowing in spare spaces. Nothing was in any particularly defined plot or pattern. All in all there was space for seven different levels to cram in plants that liked all sorts of different conditions of light and shade. Vegetable patch and orchard and wild thicket met as one. Seventy species of plant shared the available resources of light and soil and water and air in natural competition, having been chosen for their hardiness and self-seeding capability. Once again we brushed against high mint stalks, but this time blackberry prickles were not far away.

There was scope for beans or nasturtiums to climb up the trees. Apple varieties had been chosen for their fruiting across the growing season, not all at the same time. 'It should be possible to enjoy fresh fruit every month of the year,' Mr Hart told me, as we waded through the jungle, 'from the earliest gooseberries to the latest apples.' I was treated to a learned discourse on the pros and cons of the apple trees available, from Sunset to Cox to Court Pendu Plat. I was also shown pears and plums and damsons and greengages.

As with Rod Everett's garden, I doubted whether these plots could feed someone for a whole year—which would surely be their best test. All around there was an abundance of produce to suggest that the kitchen would be stocked and the stomach filled. Mr Hart gardened for his life's hobby, not as a desperate economic necessity, and no doubt some supplements came the way of his table.

For a last treat he showed me his osier bushes. He had a dozen or so, and they were coppiced every year for their long branches, which were useful for basket work. It was nearly harvest time and the fat stumps, only about a foot high, sprouted long, thin twigs over 8 feet into the air. It was incredible to think that a year's growth had made for such elegantly formed shrubs, and that they could be so useful.

As I said my goodbyes, I reflected on the amazing variety of plants and trees which Robert Hart lovingly tended. His patient and painstaking efforts were an inspiration.

CHAPTER SIXTEEN

The Centre for Alternative Technology

C roeso y Cymru—Welcome to Wales. Thanks to bees around
the border, my breakfast, which I had eaten in England,
included honey labelled 'from more than one country'.

The Centre for Alternative Technology (CAT) lay ahead in
Machynlleth, down the valley towards the sea. The centre had
begun in that *annus mirabilis* 1974, when Suma, Laurieston,
Little Salkeld and the Ecology Building Society were all opening
their doors. A man called Gerard Morgan-Grenville, who had some
good connections in the upper classes, rented an old quarry from
a local landowner on a ninety-nine year lease. He wanted to put
Schumacher's ideas about technology in the developing world into
the Western context. We need sustainable, renewable energy and
empowering technologies as much as anyone. After all, our
luxury lives squander many more resources to provide our
electricity, heat and water. The Centre has remained a co-op,
though Gerard moved on. He is remembered as an entrepreneur
start-up man and also for the fact he attracted two members of
the royal family to open sections of the centre.

In the 1980s, 500,000 visitors came to tour the exhibitions:
fabulous push-button and see-for-yourself displays; water pumped
by solar (photovoltaic) power or heated in panels of black tubing
and renovated old radiators; windmills to lift people up in

123

waterpump action. Notices in the loos explain how the sewage is composted and encourage you to pee in the plastic container to provide ready, mineral-rich fertiliser and compost activator. High up on the hillside above, windmills spin in the breeze and top up the centre's power reserves, though the whole place runs off less electricity than is needed for a couple of kettles.

At CAT there are not many concessions to the tastes of the National Trust classes. Instead there is a charming homespun neatness. It does, however, provide an exhibition circuit and a few colourful signs to mark out the helter-skelter of an organic garden which Robert Hart would have approved of, as would Sue Stickland at the HDRA. The technologists in 'The Quarry' (as the Centre is known) have had massive conflicts about where to pitch the level of their message without losing it in the dazzle of gadgetry or the superficiality of image. One of the sideshows is a political maze: to get lost in it is to find out how hard they have worked to entertain and educate. One can scramble about hither and thither on the paths or choose one's line by reading questions. The decision may take you nearer to the exit, the choices based on a good government's energy and environmental policy.

The formula of 'insight into the alternative' and 'fun day out' has kept the visitors coming. So much so that, to sustain the influx, much of the place needed renovation. The decision was taken to boost the whole enterprise and take on the many and varied tourists—from field-tripping students to family funsters, and born-again Green converts zealous for the minute details. But success hinged mostly on the mass market, the hardest to pull on issues alone.

The biggest problem, they told me, was the uphill walk to reach the exhibition plateau. It put people off, so the first priority in their £1 million share offer, needed to raise money for a complete redevelopment, was to reinstate the cliff railway which used to ferry slate out of the quarry on a water ballast system. Instead of slate, it would carry the elderly, the disabled and anyone interested in railways, up to the quarry. They would be dropped at a heritage-with-a-twist village green with shops and a café, from which the reorganised exhibits would pan out in various directions. The new layout will mean the further one goes along a route, the more detailed and demanding the ideas and explanations become.

I arrived on Sunday evening, as the day's visitors made their way home through the gardens, the adventure playground and the bookshop—all so different from a conventional tourist attraction. The climb was painless, but at the bottom I had sighed at the thought of it.

I asked the ice-cream seller where I could find Dilwyn Jenkins, the man I had come to meet. 'Try Tea Chest,' I was told. At Tea Chest, a hostel for the volunteers and a kitchen for the community who shared the quarry, a cook paged Dilwyn over the site intercom. He soon arrived with a gang of disgruntled children sick of another day at the centre. The atmosphere was relaxed as tea-time approached. 'You'd better catch me now,' Dil said, 'Come Monday morning the office is as busy as any in London.'

So I quickly claimed a mattress in the dormitory and went home with him and the kids in the back of his old Volvo. The 'hippies on the hill', as they are called by the fiercely Welsh people who surround Mac (as it is known on the hill) do not have to live at the quarry. About a third have moved to farmhouses and cottages in the district. We bundled back through Mac with Pedro, a Spanish volunteer, Dil's children Tess, Bethan and Max—all under ten—and Badger the dog. At the grocer's we picked up another little girl, some supper and more people to share it: they lived above the CAT wholefood shop in Mac's main street.

To sing for our supper, Pedro and I took the exuberant Badger for a walk. From the neck of a hill we could see the quarry windmills still fluttering to the North, but we could also see deep into the countryside to the South and West. The views were so spectacular that they lifted your soul.

Things had calmed down at the farmhouse and Dil had started to cook on the stove in the caravan outside (better than the primitive one inside). Meanwhile we played with Max and his toy car collection in the garden because he was sick of being outnumbered by ebullient girls. The peace did not last long. Suddenly the girls, dressed in disco outfits and capes, stormed the garden. A bemused Pedro was taught new phrases like 'Piggy-back, me-me-me next!' and I was dragged, kicked and screamed at to go along with it. There was only one thing for it.

For Dil's wife Clare it was a little relief. She had spent the day at the farmhouse with twenty-month old Teilo, at odds over breast milk and weaning. With a second bottle of red wine uncorked, things started to relax. We sat on a sofa in the middle of the lawn, and munched salads. The Jenkinses had been lucky to get the house, what with local objections to incomers and Welsh nationalism around Mac. Their farmer landlord had fended them off at first, until he discovered they had four children. He and his wife had four of their own and they relented. For Dil it was quite a contrast to Bristol, which he and the family had left only nine months earlier when he got the CAT publicity job. It was an even greater contrast to his South American adventures, which he and Clare recounted after supper with a slide show.

Dil had travelled to Peru's rainforest in the vacations from Cambridge University to study Indian tribes that were entirely self-sufficient. He had returned many times as an anthropologist, in four years spent courting Clare and working as a reporter for the *Lima Times*.

He had got to know one particular community. They lived a fluid existence as part of a 20,000-strong tribe, with an experienced couple acting as their leaders. These shamans held authority only by dint of the breadth of their knowledge and the respect others had for it. They were healers and arbitrators and guardians of the tribal myths. If tribesfolk didn't like it they were at liberty to move on. 'Just a single couple though, any couple, had sufficient knowledge to reproduce the whole tribe,' Dilwyn said. 'They all knew the myths and plants and drugs.' Their lives were abundant and simple, he recalled, as he told stories of rollicking evenings at village feasts; but their land became valuable and their lives were cheap. Colonists came upriver and burnt the forest to plant crops. But only coca really rooted, and subsistence farming turned into a cash economy based on cocaine terrorists. Communist guerillas from the Shining Path movement hid in the jungle and fought a rearguard attack against the landowners who sectioned off and exploited property. They had seen problems start with land rights in other countries and were making sure no one would get a toehold in the wilderness of Peru. All the same, they slaughtered and indoctrinated the native Indians caught in the middle.

On Dilwyn's return to Britain, the Green movement alone seemed to recognise the value of his experiences. He became active in Bristol as a local-authority-paid energy campaigner. He also began to organise camps for a pan-European group called the Rainbow Circle. They tried to foster nomadic ideas across the barriers of borders and languages by running camps throughout the summer, in which they brought people back to simple survival skills. The latest Rainbow Circle camp had just begun a week down my track in Monmouth on rented farmland, where tipis and tents pitched for two months, and people attended courses to learn the ways of the land by living and lying on it.

Dilwyn was right about the pace of Monday morning. It was as hectic as a London office. I just made it in time to grab some breakfast from the Tea Chest kitchens. A well-thumbed *Guardian* lay on the table. The phone never stopped ringing.

One difference between working in London and in mid-Wales was that workers at the quarry only earned £2.60 per hour. They took home salaries of over £5,000, double that of Lauriestonians, who earned roughly the same per hour but put half their time into their money-saving workshare scheme. The community who shared homes on the hill instead paid for a full-time gardener to grow vegetables, and cooked communal meals. As at Suma, pay was increased for children but, on Maoist lines, with extra for the first and second only.

Everyone in the kitchen chatted as they worked. Each person spent only a week there in summer, thanks to the rota. And there were always new people to meet. I tagged along with Roxanne, a volunteer from Liverpool who had been on the dole for six months, as she showed a couple of friends around. Ziggy, who had just qualified as a vet, and her boyfriend, had stopped off on their way to the South of France in a converted bus.

First we took the old quarry railway track and slipped through the 'residents only' gates to pick salad vegetables for the communal lunch. The flower beds, like much of the quarry, were originally pitted and uneven heaps of discarded, surplus slates. But, in sixteen years, areas had been levelled to create more space. Roxanne recalled early tales of when stone breakers chipped the slate to fill the craters—a punishment well remembered. Now craters were topped with biodegradeable rubbish like paper, wood and kitchen scraps to rot down into humus.

Lower down the slope was the vegetable garden. On the way we passed one of the two reservoirs. This one was dug into a quarry rubble crater and lined with plastic sheet. The other, higher up the hillside by the wind generators, had been built to run the water ballast railway. At that time alternative technology had needed no exhibition centre, as it lived and breathed in engineering schemes across the land. CAT commandeered the reservoir rainwater to generate electricity. When there was enough about they ran the water down the pipes to turbines at the bottom of the valley. If they had plenty, they swam in both the large and small reservoirs. But if there was a desperate call for a sudden boost to the batteries, they could drain the little one.

We dropped below a heap of slate and glimpsed the natty ladders which plunged into the black, plastic tank. Tradition had it that all bathing was nude. That was why it was so buried behind the rubble and far from visitors' view. There was enough trouble getting rid of 'hippie' preconceptions without a lot of nudity as well.

Below the reserve power tank stood the school chalets, Alpine in looks, except for their roofs of turf. Pupils on trips came to Machynlleth to learn about resources, and the centre made sure that it was not just an academic exercise. The students had to eke out the electricity and heat from the limited resources available. If one person used all the hot water, there would be no more until the sun shone again or the wind blew.

We trailed across the acre of kitchen garden along the narrow paths between raised beds. 'Don't tread on them,' Roxanne said, 'That's a cardinal sin: they are not for digging, and the gardener goes mad if they get compressed.' I picked borage flowers to add a touch of blue to the salad for lunch, with the pinch Nigel Wild had taught me in his allotment. Roxanne cut lettuces and pointed out the weeding she was responsible for. Comfrey green manure had taken hold of the plot and she asked us to uproot the nuisance wherever it appeared.

My vivid blue borage petals never made it into the salad bowl, so I had a little moan in the food queue. Ogre was responsible, they told me, if I really wanted to complain. He stood right in front of me: Peter Harper, self-confessed AT guru. Every week a senior CAT member stood as ogre—trouble shooter, shouter and literally shit-shoveller. The ogre took on every conceivable nasty job which nobody would want full-time. He ordered people around, because he was in charge, but he also composted the loos' contents.

'It's paradise here in the summer,' he told me with a look of contentment. 'But dismal in winter. Grey.' I'd asked whether there were any cracks in the façade of a well-run co-operative effort. The visitor could so easily get a rosy view. Seconds later a call over the intercom asked for an exhibit to be mended. 'I'll go round and pull it off, it'll only take a second,' Peter replied.

'No you won't,' said a grumpy head engineer who sat next to us and had his eyes buried in the *Guardian* (by now the best-thumbed I'd ever seen). 'Everything you touch takes two hours to repair.'

As a teenager in the sixties Peter had joined CND, but he went off disarmament at Exeter University, and became hooked on Third World problems. He wrote later:

'It dawned on me that the "population explosion" was part of a whole series of parallel "explosions" in resource use, pollution, urbanisation, technical change, general pressure on the environment and so on. For a theatrical temperament like mine, there was an irresistible tragic drama, and I was continuously haunted by a sense of inevitable catastrophe. Unlike nuclear war, a risk that might—but might not—happen any time, these "explosions" were continuous and self-deteriorating, unless enormous changes could be undertaken. I was at once appalled by the vision of imminent apocalypse, and stirred by the prospect of the kinds of revolutionary change that would be necessary to avoid it.'

In his twenties, Peter dressed in beads and flares and was swept along with the tide of hippies on Theodore Roszak's counter-culture.

'I was incorrigibly romantic. I believed the world could be run by hippies on peace and love.' He went to meet a pan-European group of like-minded science students, who would get together to discuss their anarchic ideas. 'We would then turn up at the local savant's and ask to be told secrets,' Peter remembered. His advice: 'If you came from far enough away and had a stylish approach you could get to meet almost anybody you wanted.'

History does not relate how the group eventually made it on to the guest list of a Nobel symposium in Stockholm entitled 'The Place of Value in a World of Fact'. Probably they gatecrashed.

'We were, in some awe, rubbing shoulders with Arthur Koestler, Margaret Mead, Linus Pauling, W. H. Auden, Konrad Lorenz, C. H. Waddington, Gunner Myrdal, Jan Tinbergen, E. H. Gombrich. But to our amazement, they didn't have any more idea of what was going on than we did. Each had a particular approach and a lot of prejudices, so there was no chance of a useful synthesis. That was the *coup de grace* to any remaining faith I might have had in Science as the Answer. From then on I became obsessed with Science as the Problem.'

Somehow Peter was became the token hippie on an international conference circuit for United Nations bureaucrats. 'There was I, this freaky-looking character who was actually quite articulate and I said: "Look there are lots of problems." And they could not answer them.'

At least they solved his money worries: 'They'd give me a wad of plane money to fly to Paris, say, and I'd hitch. Or they'd have two conferences in one place and I'd take money for both then live with a friend round the corner or under a bush.'

This hardly conformed with the ideas in Schumacher's paper on Buddhist economics, which he so admired. He began to have doubts about his romantic notions.

'If society did actually say to a bunch of hippies: 'Here's a hundred square miles, you can have it' they would have made a balls-up. They couldn't have run a whelk stall. They imagined revolution was some magic pill—once swallowed everything would change. Actually it would be even worse. The world would be smashed and return to the law of the jungle. I realised we had to have the structures all ready to move in after the revolution. The revolution would just be the enabling thing.'

In 1970, his new group planted their vegetables on their sloped roofs, constructed bike-hub wind generators, powered their radios with compost heaps studded with electrodes. Peter decided to organise his own conference. Some of his earlier experience must have rubbed off because 'Threats and Promises of Science' pulled the biggest names—including people Peter imagined would not even reply.

A follow-up event, 'Alternative Technology' was called for 1972, so named to differentiate it from Schumacher's Intermediate Technology—'too respectable and unpolitical'; Murray Bookchin's Liberatory Technology—'strongly flavoured with his irascible anarchism'—and another fad from the United States called Biotechnics—'hip-arcadian with ever an eye on how to survive the expected breakdown of society'. AT was to be environmental and political and cultural.

The most famous delegate remains anonymous. He sat in the back row and put his hand up to ask a question. 'What's all this about revolution?' he asked. 'I came here to talk about windmills.' The laughter brought the house down. Another delegate, Godfrey Boyle, came along with a bundle of *Undercurrents* magazines. Peter joined him to produce the next issue. Contributors to the first four issues printed their own work and sent it in to a Kensington address, where the contributions were compiled and dispatched in plastic bags. 'We had long arguments about whether it should be polythene, but it did allow us to have regional editions.'

Peter then went to Iran to work on the other AT, Appropriate Technology as practised by Schumacher's Intermediate Technology Development Group. Iranian villagers traditionally dug their irrigation wells above their settlements, and water seeped through the mountains to fill them. Tradition governed the sharing of water, and everyone received their portion. Along came modern technology in the form of the diesel pump. Rich farmers could sink wells and pump up water faster than it could seep to the village. So they no longer had to rely on village quotas. Their farms flourished at the expense of the rest of the villagers, before they realised the long-term harm being done. The cycle ended up with the rich buying up their neighbours and employing them. Even a windmill would have had the same effect in ambitious hands. One lesson Peter learnt in Iran was how to dress. He arrived as a quasi-hippie, with long hair and floppy clothes, and nobody took any notice of him. One day he donned a suit to see what effect it would have. Indifference instantly changed to respect.

Peter realised that he revelled in showmanship and he brought this quality to the quarry. He had arrived at CAT in 1983 and has lived in the quarry ever since. He has no plans to retire.

'That's one thing—it's hard to get rid of somebody from a community. As it is, I can't think of a better job description. I am displays co-ordinator and enjoy all the showmanship, but also all the intellectual stimulation of the issues we work with.'

His start at CAT was like a homecoming. Like-minded people worked together in a community and there was a qualified respect for some of the good old bourgeois virtues: thrift, reliability, hard work, punctuality, tidiness, honesty, deferred gratification, and rules. CAT offered scope for his enterprise, and a ready group of enthusiasts to help out. 'So, do you think that if you turned your hand to business, you would have been successful and earned lots of money?' I asked.

Peter pondered. Our discussion had continued sporadically throughout the afternoon as Ogre was called to duties around and about. 'I don't think I could ever have succeeded as a conventional entrepreneur,' he said. 'I could not do with having people work "for me". But I discovered that here there was no exploitation because everyone was on the same terms.'

We sat under the pergola in the display garden before supper. The hosepipe spurted water over various flower beds and sprinkled our feet every now and then. Peter had known Fritz Schumacher in the old days—'just an old man then, that was before he was canonised'. He rolled a cigarette and swung to survey the greenery. He alone tended it, but he hedged when I asked something about 'his garden'. 'Well, it's not mine, but...' he said.

'So, are there differences between each of you? Is there a hidden hierarchy?' I asked tangentially.

'Yes, there are differences. Some eighteen-year-old up for the summer cannot hope to have the same perspective as someone who has lived here for fourteen years.'

Things were different at the CAT when Peter arrived. He had enjoyed watching them change because most of the arguments had gone his way. Before 1983 about half the quarry team were public-school educated. The centre ran for ten years as a mixed bag of well-connected toffs and born-again hippies. But the displays had become rough at the edges, Peter found, and his Iran experience had taught him there was no point in the take-it-or-leave-it approach. It was better to buy the new suit and play the showman. 'It was as if we were standing way up on our hillside beckoning to people from miles below in the straight world to try to tempt them to look at our gadgets and come to explore our ideas,' he said, gesturing. 'I felt we had to take a few steps towards the straight world rather than ask them to cross our ravine.' So he set to work making the displays look exciting. He reasoned it was not the *Guardian's* readers who needed the AT insight but the *Sun's*. This had been his line when he backed the £1 million pound share issue, an operation called 'gear change' on the quarry. Peter had argued the case for giving visitors pure experience as against a lot of detailed information.

As realism has won out against the ideals of those beckoning on the hillside, so the community has cast off all but a couple of its public-school members and hippies. Now it consists mostly of first-generation university graduates. They have even talked about paying themselves realistic wages—in the region of £10,000 a year. 'We'll need to attract not just single people if we are to hire the expertise needed for these big budgets,' Peter said bluntly. 'It's amazing the money we have paid out to the consultants who have come up to advise us.' It was interesting that two co-ops, Dulas Engineering and Aber Instruments, which had been hived off from the centre, had increased their wages significantly since going it alone. Dulas had always made a loss in the centre's structure but had moved into the black when it went alone.

To double the quarry co-op's wages meant doubling the visitors. Peter was unabashed:

'The cliff railway will be a big pull in itself, but we don't want to be mid-Wales's number two or three tourist attraction after Powys Castle or whatever. We want to be number one. Then the two million and more visitors to the area each year will think of us first.'

Even landlord John Beaumont was pleased. Relations had been restored when he had shelved plans for a hamburger restaurant up the road. His cut of the profits would go up with the increase in visitors and there was still over eighty years to go on the lease of what had been a useless plot of land.

CHAPTER SEVENTEEN

Brig Oubridge
& the Tipi Village

I telephoned Brig Oubridge for directions to the tipi village. I rang and rang but he was never there. When finally I got through, he explained that the cat had knocked the receiver off the hook earlier in the week. I followed his directions to the letter and finally took a turning marked 'Dead end'. Both sides of the road were banked by earth four feet deep and topped with tall hawthorn hedges. I free-wheeled downhill into the evening sun. On the opposite slope I glimpsed the tipis. One blink, however and they were hidden behind hedges or trees or had sunk into the lie of the land. The road glided left and shot me to a halt in the middle of a mucky, rutted, concrete farmyard. What was left of my route was barred by three rusty cattle pens.

Two sheep hobbled off the doorstep of the whitewashed farmhouse as I sought permission to pass. There was no answer at the front door. I turned to cross the field and head where I could see activity. A stream separated us but I clambered across once with my panniers and again with the bike. I was getting used to nettle stings from visiting organic gardens.

The house above resembled a Norse longhouse, and car spares littered the track to it. A vintage Austin slumped under a bright blue plastic sheet; a write-off hulk rested with axles in the stream. I knocked at the open patio door. A woman appeared out of the gloom from behind a pool table. 'I'm looking for Brig Oubridge,' I said politely.

'This is private property,' she replied in a Yorkshire accent, and pointed. 'You'll find his field's up the hill, the one with the vehicles in.'

I had a climb to reach him and warily asked at a few tipis for Brig. Very much afraid to break the lore of the valley, I expected a sour welcome. They were off-hand, but not aggressive. Finally I found the man in his field. Brig was bent double and snipped tentatively at the nettles beside his compost heap with a pair of shears. The fenced vegetable patch looked only big enough for half a dozen cabbages. He looked up and through me while I explained myself. The curls of his long beard rippled as he swayed up, then down for a snip. I thought for an instant that he might have nettle soup on the menu, but he let them lie and offered to make tea.

We passed a cricket wicket on our way to his office in an old engineer's site caravan. We stepped inside. He and his son Kevin lived in a tipi each. This caravan housed their sink, emergency gas stove and study. They had a van, more off the road than on, for transport. 'I'm sorry I was not very welcoming when you first arrived,' he said. 'I had a TV playwright here for five hours yesterday. He was writing about life in a community and we had a very interesting discussion, but I had no chance to collect firewood for my tipi.' It was an apology for the luxury of a back-up gas stove as much as anything.

'Sugar? No, you don't look the type,' he said. I had more than a faint feeling the tipi dwellers were one up on me. I was the raw recruit; this was a gentle general with a hint of condescension.

Brig was dubbed the Councillor by tipi dwellers at the hundred-strong settlement after he was elected to the local council. He had elder status in the village, like the Vicar, an actual minister, and the Squire, the largest landowner. Some perhaps called Brig 'Counsellor' too, because he acted as the shoulder to cry on and the voice of experience. His office, a mess of yellowed, recycled paper stacked in chaos, was also the surgery.

Brig kept an old typewriter but he had bought a word processor because, he said, it was a more powerful tool. It stood by the sink, with a large Roneo machine opposite. Electricity came from a car battery wired up on the cluttered floor, and the shaky supply led to many a panic when work was in progress, because if there was a power cut he lost hours of work. Two suits hung above me as I sipped my tea. 'My Sizewell suits,' said Brig, 'I wore them to the High Court when I presented Green Party evidence at the enquiry.' Here was a fundamentalist wise in the tricks of showmanship.

His credentials were impeccable. His first tipi was the first that had been erected at Greenham Common, and had been given to him by a friend. His present one was a veteran of Molesworth peace camp winters. The office cabin, fitted for revolution, also housed the telephone line from which a cable led to the tipi.

'I have no doubt it is bugged,' Brig said, 'and that MI5, or whoever, have files on me. I've been on the UK Green Party Council for five of the last eight years and on CND's National Council for another two.' I was inclined to believe him. The Councillor and his merry band were a tough and mean-looking bunch with beards, tattoos, and leather trousers. The band were pitted against strong forces, if not of the state, then of the elements.

I asked for some basic tips on etiquette before I put a foot wrong. This was a tribal society, though only an hour's drive from Swansea. 'Always say "knock, knock" or something when you visit a lodge,' Brig said. 'And always take your shoes off before you enter.'

I drank up and Brig dispatched me to Big Lodge, the communal wigwam, where Silvano was looking after the visitors. I arranged to return the next day. 'I'll be here to listen to the India-England test match. I'm supporting England,' Brig said, earnestly for an instant. Brig the sportsman had supported Middlesex until Gatting toured South Africa.

I strode the highway, panniers slung over my shoulders, the cowboy stranger in town. This really was frontier land, with everyone ready to move if the call came and, I feared, despite my confident gait, ready with a scrap iron armoury to see off any unheralded surprises. And their dogs would howl.

I stopped by a lodge, its tarpaulin grey and stained brown about its chimney crown. The bracken had worn away from the skirt suggesting that it had been pitched there for months. 'Hi, looking for Big Lodge,' I said.

'Show yourself,' a gruff Australian replied.

I leant down to his doorway and he pointed my way without moving from his cross-legged seat by the hearth. Twin girls about six years old, dressed only in sweatshirts, squabbled. He, seemingly their guardian, strained tetchily. 'It's the biggest lodge you'll see, in the bracken over the stile.'

Big Lodge was big indeed. On one side of the west-facing entrance was a stack of logs, on the other a shrine to everything ethnic: a plastic buddha, a carved African death mask, a forked branch hung with bells, squirrel's tail, beads and yellow ribbon. 'Knock, knock.' I took off my sandals and walked in, without having to stoop for this tall entrance. Inside it was spacious and atmospheric. I introduced myself to the four men who lounged by the fire on sheepskins, then knelt on the rushes. The Welsh had improvised the floor out of marsh grass, according to Brig—the same stuff Bruce Marshall fought back on his peat bog.

Silvano, a huge Italian who made me think of a character out of

Eric Newby's *Love and War in the Appenines*, huddled by the hearth of flat granite rocks. He stirred a broth over the fire on a three-legged iron stool with home-forged utensils. Another iron stake dangled a chain from which to hang the cauldrons. Lenny, an Irishman, strummed his guitar in the background. The others, new arrivals Mark and Graham, sat in silence mesmerised by the fire. All but Lenny looked the part, I thought: waistcoats, loose shirts and trousers, all patched and hand-sewn to remedy wear and tear. Silvano and Graham had beards, though neither was a patch on Brig's tumbling wizard locks. Graham also wore John Lennon-style pebble glasses. Lenny dressed more conventionally in a green pullover and trousers. He wore his hair short.

I felt an intruder as I broke the silence, intrigued to hear more about the valley before I quizzed Brig. I explained my quest and they all asked me how long I was to stay. I had a couple of days. This, they said, was not enough for a really in-depth picture.

'I've found I only get more confused about my impressions when I stay for a long time,' I defended myself. 'It's the people I'm interested in most, their psychology and how they came to seek this alternative life, as much as what the place is like. So, if I can get chance to talk to a few people who have experience and a long view I go away with their direct impressions and a clearer, more objective picture.'

That only served to silence them. A moody night was developing outside and a cosiness around the fireside only brought more introspection. But I discovered that Mark knew Roxanne and Ziggie whom I had met at Machynlleth. He had studied philosophy at Liverpool at the same time that they were there. Mark had since worked for eight months but had now decided to tour and take things as they came. 'Perhaps I'll go south for the hop harvest.'

Graham was a sociology graduate about the same age. He had been at Essex University, where everybody had told him he was twenty years too late. He had stuck the course, he did not know how, and now shrugged off the idea of conventional education with a studied disaffection. He had wanted to be a professional chess player but went to live at Wheatstone commune instead. By his account, the place was in crisis and starting again from scratch since the last generation of communards had dispersed. His new goal was to farm 'veganically' when his communal group managed to secure the rights of ownership to the property.

Graham and Mark briefly struck up a conversation about Foucault. Mark gave a synopsis of his thesis on repression in Western thought and particularly the supression of feminine ideals. Briefly, the disaffected Graham showed an animated fondness for the intellectual. Lenny strummed and sneered at this obviously bourgeois company.

Silvano erupted. For anarchic or nosey reasons he had fumbled through Mark's bag and had found a cut-throat razor and stone. He brandished it. 'This razor?' He asked in broken English learnt from three years in the valley. 'There thirty children in tipi village. Come here any time. Thirty children. You not put this deep into bags. You not close case. Razor very, very sharp. Thirty children.' Mark was at a loss for words nor did his apologies placate the Italian who stood before him and shook his head. Silvano did not understand the evasive reply or chose not to. He turned away with mutters and the stilted conversation again died a death.

Graham decided to make chapatis—a skill he had learnt on the Kaftan trail. 'But,' he apologised as he kneaded wholewheat flour, 'I haven't managed to get the light spongy texture they have out there.' Even so, he had brought his own miniature rolling pin and wok for the display and everyone tried to be suitably impressed when we shared the pile with Silvano's broth. Supper was ready and it was open house for all. Silvano insisted that he personally serve each of us. Tea appeared from the food chest—a feature of all lodges, to keep cats and rats from fresh supplies. Water from an adjacent bucket was ladelled out by coconut shell and boiled. Graham strained each cup of his own blend of green leaves with another coconut bored with holes.

Two German visitors, Klaus and Corrina, had arrived, and they wolfed down their supper. Germans often called, Brig had explained, because he once wrote the Welsh chapter for a German friend's book about Green Europe. 'So, are you foreign born?' I had asked, wondering how he was christened Brig.

'No,' he had replied. 'Brig is just short for Oubridge.'

Corrina had come to the valley the summer before and Klaus had cycled from Germany, via Greece. I opened with some bike talk.

'What sort of machine do you ride?' I asked Klaus. He strained to understand.

'Bicycle, yes.'

'How many gears do you have?' I said, conscious that my ten could have been doubled to spare the grind up the hills between Machynlleth and Rhayader.

'Three,' Klaus answered, with his jaw set tight. I thought he had misunderstood me, but no. 'It's enough,' he said proudly, and shook his curly fair hair out of his eyes.

Corrina unrolled her bedding from one of the bundles around the edge of the tent and asked to borrow the oil lamp. It was Silvano's. 'You need?' he asked frostily. 'You need, all need. You take.' He shrugged his shoulders. Corrina took the lamp unflinching and read *The Magus*, a compendium on the occult. The air cleared, with

Lenny strumming as Silvano sang a mournful Italian chant.

I took out a pair of shoes and socks from my heap because the evening had chilled. I loosened the laces. 'What are you doing?' Lenny shrieked. Again the atmosphere froze. I realised that I had broken the taboo on wearing shoes indoors.

'Not in here. Socks yes, but...' I decided that the best remedy for the smoke and the glare was sleep. So I unravelled my sleeping bag and pulled out extra warm clothes. In seconds I was dead to the world.

My sleep was fitful, though I had plenty of room, slotted between Mark and Klaus. The mangy camp kitten, that had hopped adeptly over the hearth stones during the evening, found my makeshift pillow of tee-shirts a cosy nest and dozed there all through the night. I tossed and turned to shoo the animal away considering the likelihood of fleas. But every time I woke the kitten had crept back.

Silvano got up at about nine and rustled up porridge from the food chest. It was a damp drizzling day outside. He stoked the fire and crouched over it dressed only in a checked shirt. Graham and Mark also appeared, and stretched. 'What you do here?' Silvano asked Graham abruptly.

Graham was lost for words. Was the question another prickly Italian trick? Silvano was conspiratorial with his movements and everyone visibly quaked if they had to answer him. 'Next time you come you sit in the corner and watch everything other people do. You sit quiet. You watch,' Silvano said. Everyone stayed on their guard but Silvano turned to attack the porridge.

Graham rustled around for his green tea-leaves to brew a morning cuppa. They were nowhere to be seen. Klaus, sleeping like a baby, stirred with a devilish look. I assumed he had mistaken them for something more potent.

We ate the porridge, rich in currants and dried fruit, and prepared to leave. Silvano said, melodramatically: 'Remember this tipi has two doors. One you leave, the other enter into the world outside. Always prepare yourself. 'I have a tipi now made. I will travel but it is not me will take my tipi, it is the lodge will lead me.'

Lenny had heard Silvano's patter before; he left in time to cadge a lift to the dole office. Graham and I chopped our share of the firewood then strolled back to the track, I to interview Brig, he to walk to a station. He was on his way to Hull for a meeting with the Birmingham's Radical Roots and New University folk. They had teamed up with anarchic co-ops across the North—with Zebedee's Café from Birmingham taking the lead—to plan expansion strategies, and Graham's Wheatstone co-op would be their countryside training base.

I went to interrupt the cricket. Brig described the valley as a university of outdoor living, peopled by professors, researchers, postgraduates, undergraduates and freshers. Most freshers passed through. Some stuck at it. A few graduated, made the village their home and carried on their own research. All sought the answer to living in harmony with the environment on a basis of minimum consumption. Live simply so others may simply live.

We sat in Brig's lodge with the cricket on the radio in the background and chatted until the commentary reached crisis pitch, when I was made to hush. Sheepskins insulated the cold ground. There was a food box, the customary stone hearth and the only telephone in the village. The greyed canvas cathedral above reflected the dingy day outside. Brig and Kevin could have done with a woman about the house, I thought, and was struck by the sexism of the phrase. I sipped my tea quietly as Brig held forth.

The tipi always confronted you with the results of your lifestyle. Not without smokey reason was it dubbed a glorified chimney for living. To cook you needed food from the land or the nearest shop, firewood and water—all carried from the road or husbanded. All rubbish had to be taken back eventually. Consequently the roadside plots were at a premium. Brig lived near the path, whereas I had stayed much further off amongst visitors and the undergraduates, and the atmosphere was a good deal more fundamental as a result. They tested newcomers' commitment by the distance.

It was one of the assessment tools which had developed since 1976, when the community first erected a tented camp. Billy Busk, the local farmer, had found that his land fetched more sold in small parcels to hippy campers than for agricultural use so he decided to sell up. There was a market for his plots in the convoys that wintered on common land across Wales and a Lampeter-based group bought up fields.

The local occupation really mushroomed when Busk began to sell fields on hire purchase at £5 a week and the next-door landowner Captain Blunt, a former navy officer complete with parrot, joined in. The buyers bought title to the ground but none put up permanent buildings so, Brig felt, a certain equality remained. As time went by and old hands sold up, land was secured in a trust to retain fluidity. Never bricks and mortar, never 'property as theft'. Brig, the Councillor, stood as a linchpin to the common land system, but he was not a tipi pioneer. He was not one of the first to pitch tent, nor had he leapt in at once when he came upon the settlement.

In the late sixties he had studied civil engineering. He had spent a summer vacation crossing the United States, and had returned full of the joys of radicalism. His college was not so impressed, and Brig was thrown out. The time was right for India and the kaftan trail. 'But the trips were extremes,' he recalled 'I had a personal handle on the realities of the fact that unjust, conspicuously affluent consumption didn't lead to a more balanced lifestyle, and that the simple life was happier.'

Brig had sat it out in simplicity, odd-jobbing from a London squat until 1977 when new laws squeezed him out. Others took to communes or caravans; he took to a van and ended up in Lampeter running a market stall. Through the Lampeter connection, the lodgers in Talley had asked him to open once a week on their hillside. He brought second hand clothes, matches, candles, rice and other provisions. It became a regular thing and in time, Brig was given a tipi. He bought a share in a field in 1979 and set up shop in the valley for good.

The election of 1979 had proved a turning point for Brig, not because of the Tory swing but because of the Green Party. 'Previously an unknown club of friends in a pub back room', they had fielded fifty-three candidates and so had squeezed into the party political broadcast bracket. 'They were the most significant force in the election,' Brig told me. 'The majority only thought of the turn and turn about with Labour and Conservative, but this was altogether more exciting.'

Almost overnight the party's membership rose to 3,000. There was a certain amount of middle-class dreaming which jarred with Brig, but he was inspired to join. He was, besides, already out there living his politics. He had actually done it the hard way, on his own. He had changed himself first before he decided to preach to others. That gave him the edge. He could not be called a hypocrite easily and all through the Thatcher years he stayed out in the cold. So, seemingly effortlessly, he had risen to influence as a voice of experience.

Meanwhile the shop enterprise remained his livelihood and a focal point to village life. Then, in 1984, Brig was asked to present the Green Party's evidence at the Sizewell nuclear inquiry. It meant three months off, so he handed the shop over. 'It was infuriating anyway,' he joked. 'Every time you got stocked up, all these customers would come along and buy it all.' The never-ending round had at least give him access to all the village gossip and he had earned respect as a listener across the counter.

His experience had taught him that the problems of a simple life were inextricably linked to those of society on a macro-level.

'We, here, are about as far away as you can get from the wasteful, polluted, consumer society than anywhere in Britain. Yet we still get buzzed by military jets 50 feet above our lodges. On that level dropping out is an illusion. So dealing with your own life is not enough. You are linked to society and must deal with that.'

I could see his point. Even in the deep woods and valleys I had passed on my way through Wales, where I could have lain all day on the roadside without seeing a soul, there were postboxes and telephones. 'So can you imagine the whole population living out here?' I asked.

'No, I don't expect them to give up house and home to live in a tipi,' he said.

'That's a totally unreasonable expectation. But the more people who gain an experience of a simplified lifestyle, in a caravan or bumming around the Third World, the better in terms of their assumptions. People can then make up their minds and bring themselves to a more ecological point.'

I joked that he and the other members of the tipi university ought to be running paid-for sessions on Gandhian politics or courses from *The Tipi Book*, which had originally inspired them. 'People can turn up and do their own course depending on who they bump into,' was the reply. 'That's a much better way of going about it. It would have the same effect but would not cost £200.' He reckoned that Talley Valley residents would never be organised enough to get a proper course going. Courses were the sort of thing run by middle-class communities who bought their way into large houses in the country. 'Quite élitist.' The valley acted as a safe haven from money-based systems for all sorts of people thrust over the edge in Thatcher's Britain. No one paid for a night in Big Lodge and no one gained from giving hospitality.

'The squatters of the seventies made a decision not to go straight. Now the legacy of the Thatcher years has created an underclass who never had a route into conventional society. They come to the valley out of economic desperation.'

As if to make the point, a villager popped in to ask if he could use the phone to call the hospital. A pregnant woman had walked into the camp the week before 'as her last resort' and he had got her medical attention when she had started labour. Now he was going to collect her.

'She may have been in a state when she arrived but we did do something to calm her down. She would always have the option of a home with us. Valley life makes people realise they are not helpless, because it does not take long to build or buy a home here. So you can see the change by your own efforts.'

To see this man take the time to care for someone he had known for only a week was heartwarming. He returned to pay for the phone call (Brig never managed to cover his phone bill even though he always got cash in advance). The woman had discharged herself, her baby stillborn. The child had been dead inside her for some time, the hospital reported.

This sort of act changed my view of the camp. Before I had considered it as just a scruffy pack of outcasts who lived the anarchic way amid grime and litter and scrap metal. In fact, you needed a stoical commitment to pull through. The young man who had telephoned the hospital had lived on the site nearly all his life. The Green ethos of empowerment, then, pervaded the life of all tipi settlers.

Fewer people were working by the late eighties, Brig said, though some still practised crafts, drawing enterprise allowance. It was possible to live on state benefits like single parent allowances and still legally own a tipi. And, Brig reasoned, the state's burden was actually lower because with a tipi they had few costs to pass on to the DHSS, whereas if they lived in town their rent subsidies would be astronomical. 'The fact is, though,' I pressed him, 'you can only survive with an umbilical cord to the state where people work and pay taxes to fund you. So your life is as dependent on the success of other people's capitalism as any other.'

Brig's sensitive approach became wistful. 'We've got to sap the taxes because the state is rotten anyway.'

But here was an issue I wanted to explore. The community had been formed to live simply. They needed some equipment to break away but most of it was only affordable because it was cast off by the society they had rejected. 'If someone gave you a million pounds,' I asked, 'what would you do then? Would you become a consumer like the rest of us?'

'I don't need much money. I'm not a big spender,' said Brig. 'If I had more I would spend more—and I wouldn't want to because that would detract from my ideals.'

I puzzled over the sincerity of his point and could reach no conclusion.

'I could enjoy a new valley mini-bus for shopping. The old one is falling apart. Then there would be a debate about whether a brand new one would be better ecologically since it would be fuel-efficient and cleaner. I tend to think this techno-fix doesn't work because it would probably take more energy to produce the thing in the first place than all the mess we cause by burning fuel inefficiently in a beat-up old van.'

This seemed fair enough, but it was still a nibble out of the wider society's 'have your cake and eat it' way of consumption rather than a complete alternative. The valley had cars and motorbikes. Brig himself owned one, and a computer and a telephone.
'The technological process is in control of society not the other way around,' he said and I agreed. He had admitted that self-sufficiency outside that society was an illusion, too. So what was the answer. Was it really worth the punishment he inflicted on himself to step outside society at all? We were all trapped anyway. 'Either you get out or get involved, I say. Either don't be plumbed into the standard sewage system, or else get involved in collective action to block the authorities pumping sewage into the sea.' Brig himself was not plumbed in. His loo was a bucket deposited safely on his compost heap. Robert Hart and David Stephens would have been pleased. Brig was not plugged into the National Grid for electricity either.
'If you live by habit and not question, you are making no progress,' he said.
'And your car?' I asked.

'Yes, it is still unsatisfactory. There is no adequate public transport near here and the nature of my work has forced me to make long journeys. But I try to keep my own car off the road for two or three months every year to prove to myself as much as to anyone that I'm not dependent on it.'

The benefit of life in a community was that they could share. Most weeks the hundred residents in the valley survived with just six or seven cars making two trips a week. 'It's a step forward, though not the ultimate one.'
I returned to the idea of inconvenience. Was it inevitable for major change only to happen when people were not inconvenienced? Brig quoted Ivan Illich. Illich had studied car drivers in the USA to find the average mileage of each car each year. Then he worked out the time the owners spent driving and cleaning them, as well as the time they spent earning money to pay for them and for repairs and petrol. It turned out that if all this time were taken into consideration, they had really only travelled at four miles per hour—a brisk walking pace. So the convenience itself is an illusion. You are forced to make sacrifices

elsewhere in your life. You work to buy the car, to run it and keep it. You are trapped.

'And you still have your computer made from society's drive for technological advancement, and the circle of work to pay for it?'

'Well, yes, once you take part in politics and communications, you take part as it is. It is computer-addicted.' This was fair enough, but Brig had also said how the Green Party had been hijacked by centralists and power-brokers in London who had sacrificed ideals for a stake in the action. They had decided to 'be like them to beat them'. Wasn't that what he was doing?

'Life is about making the most appropriate compromises. Only when you start to set yourself up outside do you see the conflict of your links. You are only Green with a shift in consciousness. We have to look at the energy behind, that powers the technology, and sort it out on a society level before we use the tools themselves.'

The most important thing you could do was make choices about which connections you made to the consumer society or the technological society or the resource-wasting society. And by way of constructive suggestions to add to this politician's answer, Brig mentioned a Rainbow housing co-op in Milton Keynes where the street shared laundry facilities. Yes, they accepted that labour-saving technology was liberating, but it was most appropriate in a community sense. There is no need for a washing machine and dryer in each house, when they are used only three times a week. In a laundry they got to know other people and also shared the resources which went into each machine. They started to see that they could make choices in their lives which didn't trap them into consumerism. The same could be said for lawn mowers, for cars, for bicycles.

Brig's mail arrived. Kevin brought it in when he went to pick up his friends to play cricket. He called his father Brig and asked politely if he could have some food. Brig was forthright and categoric: he had eaten most of the biscuits and Kevin could only have a reasonable share. Once they were gone, they were gone.

One letter was from an insurance company. Brig had been so irritated by their continual junk circulars that, he had written to request a quote for his property—one tipi, constant fire risk, impossible to lock. Broking Officer P. Simons (Mrs) replied to the effect that no cover was available. I asked how Mrs Simons, for instance, might change her world and pick up on the Green thoughts she'd perhaps had, prompted by magazine articles and supermarket shelves. 'Oh, she ought to leave her husband and start again,' Brig said wryly. 'I've seen a good few women do it. She almost certainly wouldn't want him. He's probably more trapped than she is.' Brig spoke with a glint in his eye.

The eyes lit up again when I offered to help him catch up on the time I had taken out of his day. His rush matting was past it and I could cut a sheaf or two to replace it. He showed how I was to thrash the stems to make sure the rushes seeded for further harvests next year, handed me the bread knife—best tool for the job—and disappeared down the hill for some peace and quiet with the radio.

I caught the Camarthen train, Cheltenham-bound for a weekend with my girlfriend Clare. It's a funny thing about travel—going abroad, you cut yourself off by language, by culture, by currency. A tour of Britain has telephones, invitations and lazy weekends.

We drove out of Cheltenham with Mr and Mrs Jones, family friends of Clare's parents, for Saturday lunch in a Cotswold pub. Philip Jones ruminated on catalytic converters as he drove. He had pondered fitting one and explained to his wife how it worked.

I played the young radical: 'Sell the car!' To which Mr Jones replied:

'Of course, your generation have it different. I had my first car, a Sunbeam, when I was thirty, and a bicycle until then. And my father would put his car up on blocks for the winter because he never drove it for three months of the year. It was common practice then, because it was too cold to drive without heaters. You could pay the tax by the quarter, you see, and everyone saved for three months. You lot have cars as soon as you can drive.'

A summer's afternoon in August. The weather was hot—was it the greenhouse effect? We had the luxury of air-conditioning. I looked forward to lunch and decided not to start an argument about the CFC content of the air-conditioner.

CHAPTER EIGHTEEN

John Button, Green author

He's a stirrer, is Brig. We need more of them in the Green movement,' John Button said when I told him about Brig's letter to Mrs Simons, the insurance broker. He was shuffling a sheaf of rail tickets and invoices on the dining table, sorting out his business expenses since his move to Stroud earlier in the year from Kirkcudbright, Galloway. His new terraced home looked from the outside like a one-up one-down tacked on to the end of a back street. In actual fact it was a warren with three rooms behind and more on top. Each had either a shelved wall or a toppled stack of books.

John told me another Brig story as I drank apple juice in his tiny front room. Once upon a time Brig Oubridge had written to the Ministry of Defence to ask about the intricacies of their Trident or Cruise policy. A letter came back by return of post, which the old peace campaigner thought was highly unusual. Still more out of the ordinary, he discovered that the Minister himself had picked up his enquiry and invited him to meet at Whitehall. The letter began 'Dear Brigadier'. Mr Oubridge replied with kind acceptance and pointed out the mistake. The appointment was duly shelved; something had cropped up.

John knew just about everybody in the Green movement, it turned out—Jonathon and Sara (Parkin) and Brig. They worked very much together on a national level, no matter what they also did on a local scale. John was a communicator, not a politician.

He had written *Green Pages*, listing everything you could possibly imagine with a Green tinge to it, up to and including Exchange Resources, a Bath-based job agency recruiting for the computer industry on the ethical requirements of the employee rather than employer. He had written a *Dictionary of Green Ideas*, a *Green Guide to England* and the popular *How to be Green*. His friends in high places had once asked him to put together the Green Party manifesto, which he duly did, even though he was not a member. They signed him up swiftly.

It was no surprise to hear that there were more strings to the man's bow: from publisher to author's agent to editor. John had been the publisher at the Findhorn Community, near Inverness, the biggest commune in Britain. His task of the moment was to edit the New Consumer books from Richard Adams's Newcastle think-tank. But John also led regular workshops on the psychological side of empowerment and how to relate to other people (he worked in group therapy sessions with the Green Party Council on their committee structure and rationale). He was the first town planner in Shetland in the seventies, appointed some twenty years after the legislation allowed for one. Most significantly, he knew what 'everyone else' was up to, how they got there and what their latest ideas were. He was a trustee of the Eco Environmental Education Trust set up by Matt Dunwell and Pat Fleming, and a consultant to Green Books.

There was something very special about John's awareness which attracted me and made me realise what I had blindly sought throughout. Green ideas could best be promoted if one can understand why people refute them. This awareness became apparent when John described Findhorn. He had been asked to join the community as their publisher after he had co-published a book with them on Celtic art. The community, sometimes called the University of Light, was by then well established, with over 200 members.

'What I find hard about the place is that I don't think they have much of a political critique. Certainly until the time we were there there was a real fear of politics because Findhorn sees itself as a spiritual community. I think it's vitally important to know how you relate to the world in terms of power structures. Who has control of what? Who has power over who? It's not seen like that, which is a very privileged position. Basically, privileged people don't need to look at who has power over who, because they have got it.'

Nonetheless, John lasted four years there—longer than average—

and praised their work. It is a hub for a great many people. He learnt massage, psychosynthesis, *Gestalt*, co-counselling and body work. He met the likes of Fritjof Capra and Marilyn Ferguson, and he started to lead group work himself. 'Do those sorts of things have a relevance?' I asked, acting as devil's advocate. John had so clearly grasped the critique of his position that he did not rise to the bait.

'They do seem to be more "fringe-y" than the tendency who say: "put your rubbish on the compost, not in the dustbin." To most people running round naked in the sauna and going for a massage is more alien to their Anglo-Saxon nature than going out saving bottles. I'm just a great believer in "whatever works". There were two events alongside each other at the Green Show which I helped organise recently. Mind, Body and Spirit was utterly packed out while the Green Consumer Exhibition was about half full. So when you say that many people aren't interested in massage, I don't think it's true. I do think it tends to be different people.

'I am very interested in bridging the gap so that people can see that they are not separate things; therapy and healing and spirituality aren't something different to recycling and rambling and wildlife management and so on. They have got to be two aspects of the same thing. I saw the need to combine them when it came to *Green Pages*, for instance. I could see really clearly how the whole span fitted together and there was not any bit that really stood out. Some people said: "Why have you put this in?" But in general the overwhelming reaction was: "Really good to see everything together." '

'I think everybody I have met has kept a very whole, holistic view of Green as far as their philosophy goes,' I said. 'But they don't really want to use the word "Green" except in terms of the exploitation of the environment because it has been taken over and now it suggests middle-class concern with the land.'
'I think it's important for us to take it back,' replied John.

'But the obverse of this is that nothing is sacred, nothing is unquestionable. It's just as important to ask questions about spiritual positions and healing techniques as it is about political campaigning.
'You will find that in every area of the Green movement there are people who think what they are into is it, full stop. That's fine as long as they are open to question about it. Because we live in the real world, if we can't form a critique, like a Marxist or a fundamentalist, in relation to other positions, we tend to become fundamentalists

ourselves. I think the most important things about the Greens' view is that is eclectic and it listens, it knows where it stands, but it's not closed.'

In 1983, John and his wife Margaret decided to live separately but close by. John moved to Laurieston Hall when it was still a commune. Everyone lived together in the house. This was different from what I saw. The housing co-op I saw was more like the Findhorn ideal: they lived separately, but could choose communal living when they wanted it. John put his Laurieston experience down to 'character building'. He arrived at a time when all hands were needed for repairs and the Hall's upkeep but he had joined on condition that he could carry on freelance work outside the community.

'You have almost always got one person who is the butt of current criticism. It may well have something to do with them, but quite often it does not. I just became that person for a while. I left before they threw me out. Since then, as you know, Laurieston has stopped being a commune and become a housing co-op. I think they had just chosen to side-step the issues, which is a perfectly valid way of doing things. Either you deal with them, or you decide not to deal with them. At the time I was there they had not actually chosen to do either.'

The focus of John's work has changed somewhat since this time. He has lived alone and worked freelance, 'committed but not centrally involved in any one campaign'. He had obviously developed a fervent need to square the political and relationship circles. Specifically, he feels an ethical framework is central.

'If you don't know where you stand on human rights or environmental rights or whatever, you tend to get swayed by whoever spoke to you last. But you don't have to have a rigid creed, I do think it's possible to have flexible positions. For instance, if you see a child being molested and you know it's not OK, you can do something about it or you can choose not to. I am not saying you must go in every time, simply that you must know whether or not you think it is right or wrong; if you don't know where you stand, you won't know whether there is anything to be done.

'On a smaller scale, we have a litter problem. You see someone drop some. If you don't have an ethical stance, you would not even think: 'That's not on'. When you have an ethical stance you can choose what to do. There is no rule: 'If you see this happen then you should...' What you do is acknowledge that this is a not-

OK thing to have happened and therefore something must be done based on what you think you are capable of doing. So, if you think it is the appropriate thing to do to say: 'Hang on a minute, you have dropped this and I think it would be really good if you kept it rather than dropped it down', then you do it. If you don't, then you pick it up yourself. But what you don't do is nothing.'

I followed his line of thought. But I felt that this attitude had not reached the mainstream yet: that people were good at complaining about a problem but not effective in acting upon it. So how did one bring about this change in attitudes?

'It seems obvious to me the change needs to happen in a radical way—and radical means deep, revolutionary, far-reaching. Also I don't think you can force change to happen. You can set an example, but it's very much like the Californian light bulb joke. How many people does it take to change a light bulb in California? Only one, but the light bulb has to want to change. That's true with people. They will only change when they want to change.

'I also believe that along with the ethical stance go Green articles of faith. One of my Green articles is that basically people are intelligent and know what needs to be done. The only reason, and it is a very big one, why people don't do what needs to be done is that they don't have full information, they don't have a full range of courses of action, and they have been damaged in the past socially, emotionally and mentally.'

But surely there were plenty of people with vested interest in the status quo. Many successful capitalists would argue that they were well aware of what they were doing, and justify their exploitation of resources.

'Well, of course, but another aspect of the articles of faith is that I believe very strongly that oppression does not help anybody, neither the oppressor nor the oppressed. It might provide a short term advantage, but in the end it does not actually benefit anybody.'

I did not agree. I felt sure that a lot of oppressors were quite happy with their swimming pools in the back garden and tennis courts, and bank accounts which meant they didn't have to worry.

'The trouble is,' countered John, 'if you believe that, then you believe that they are getting happier by being more oppressive. I can't see a solution except in what Marx said: "The privileged will never give up privilege without revolution." My experience is that beyond a certain level, in general, additional material privilege does not make people happier.'

'Security is what makes people happier. Isn't material security real?' I asked.

'What do you mean by real? I don't think that's automatically true. Material privilege, as it amasses, often makes people less secure, which is why they feel the need to be more and more protected from the outside world. That's something really worth exploring.'

I was led again to consider the notional individual I'd discussed with Brig Oubridge, the person with some sympathy for the Green movement, uneasy about the state of the world, but unaware of their own impact—the man or woman selling insurance. Does she or he care about the false security being sold—or security that may impoverish the client?

'I would be really interested to talk to the person and say: "You want to offer me insurance, but what is it you are actually offering?" I think people can get much further much faster trying to answer the questions themselves. It's just that they may never have asked themselves what insurance is, they just sell it.

'People call ecology the subversive science, and I think Green is incredibly subversive in so far as it gets in underneath the system and eats away at it by asking difficult questions. That's partly why I was interested to talk to the Advertising Association at their annual Peterhouse Seminar. This year I was invited to lay out the Green agenda—because there increasingly is a recognisable agenda to follow.

'So I was able to ask questions like: "What is advertising? What is it doing? Who are you benefiting?" They were interested to talk about all the very fundamental questions. I think that's one of the biggest changes in the last few years. People feel very much that they don't need the same loyalty to the cause that they needed before, especially in the present climate of change. So the insurance person, I think, is much more willing to talk about insurance and security in the light of a changing world—as long as he or she does not feel threatened.'

It was lunchtime. We strolled down to the shops in Stroud and headed for the usual vegetarian café with its lively paintwork. Being Gloucestershire, this one had a more upmarket feel and art that was for sale adorned the walls. We talked Green Books. I asked advice.

I had regarded John Button as something like a star, though his look was somewhat unlikely. He was dressed more as a folk

musician and carried a purse like my mother's. I did not know
what to make of it or the shopping bag that dangled from his
forearm in the Post Office. John even admitted that he got fan-
mail. The best antidote he knew for any ego trip was to 'get out of
context'. He may have been a Green guru at the party conference,
but at a music festival he was nobody.

I remembered the Peterhouse Seminar, with all the top
advertisers. Didn't the eyes of the top people light up as they saw
new ways to sell everything? Did anything really change? Had they
not just listened to what he said and twisted it for their own use?

'It's tricky, but I will talk to anyone who wants to talk. We can't
ignore people with power.'

CHAPTER NINETEEN

Misha Norland,
homoeopath

Misha Norland could have been an opera singer. He was stocky and round-faced, his smiling jowls covered with a greying beard. He was, in fact, a homoeopath. The School of Homoeopathy at his home was the first such institution I had visited.

We chatted. I probed Misha's story, he probed mine, which was apparently typical of a homoeopath's first meeting. Then the practitioner (not 'healer', definitely not 'doctor') would take over and delve into my problems with a view to identifying patterns of disease. Misha went further into my life history because his particular development of the homoeopathic craft was to combine the more empirical philosophies of homoeopathy with those of psychotherapy and alchemy. If my life were a river he would look at how my health, the water, flowed, how it ran smooth between banks, how it had buffeted the physical and mental sides and how, with illness, the water had been dammed at the banks. 'In the wake of dammed-up areas, where the stream of water does not flow, stagnant backwaters, places where disease may flourish, form themselves,' he said. Misha seldom used laboratory procedures in order to probe—such methods only observe the effects of the disease, the products of sickness. He was more concerned with identifying the subtle factors that obstruct the flow.

'It's the same with the environment,' he said. 'The ozone hole is an end product. It's not much good studying that to combat the problem.' Misha framed his theory thus:

'It is understood that ill-health is the result of diminished or disordered vitality and the symptoms of the disease are the language used by the individual to express the uniqueness of the disturbance. The healer's art is directed towards the establishment of the self and vitality through a process of awakening awareness. Health is defined as the organism's ability to adapt itself to circumstances, revealing latent creativity in terms of freedom from rigidity and fixed habits of body, emotions and mind.'

Clearly I would get no panacea pills. Misha was interviewing me, not the other way round. He listened to my questions and explained his *modus operandi*, but skirted a full-blown discussion of his deep motivations. Rather, he fended off some of my questions with asides to his children. The boys roamed the beautiful gardens of Yondercott House, near Uffculme, Devon, and filled the air with the sound of play. Every now and then they would burst in with a broken bow or arrow to mend, perhaps a few shrieks of jealousy. There was always the Lego to rattle across the floor in the room Misha and I occupied, and my saddle-bags to unpack.

In childhood, Misha remembered, his responses to sight and smell and sensation were at their most 'visceral'—gut reactions. In treatment he looks for moments when we likewise reveal our most essential soul. At the instant of alarm, perhaps, the Italian waiter in London clicks back into talking Italian. Misha would focus on such painful moments of my case history to see my reaction.

'There will have been some events in the past which were very intensive and which, in order to survive the impact, required you to adopt a survival posture. I am particularly interested in how you reacted at these moments. You see, the healthy organism would re-establish its equilibrium after the exciting event had occurred. But in sickness this 'survival posture' is maintained beyond the duration of usefulness—it becomes the obstacle in the flow of the river of life and leads to the intensification of suffering. Under the action of the homoeopathic stimulus these traumatic events are recalled, and after their recollection and re-enactment—pain, tears, fear, anger, etc—detoxification occurs. This is the process which is universal to all healing therapies. So it's often said that symptoms are aggravated prior to a cure.'

The apprehension of being in the hands of a homoeopath was a new feeling for me. I lost my considered view and reverted to blacks and whites. I wanted to know what I had that might kill me. Would I suffer painfully? Could I prevent it with a healthier life?

'Of course you are going to die,' said Misha matter-of-factly. 'Let's look at it. You say you wish not to suffer. What's this idea of not suffering?'

'I don't know,' I replied sheepishly, 'I was thinking of cancer...'

'Why would you like not to die?'

'I would not like to die prematurely.'

'What is it about prematurely...?'

'When you think about your life, you plan to four score years and ten. You consider...'

'Why do you plan?'

'Because you think you'll go that far.'

'...Because somebody said that, because there is an expectation. Let's look at the expectation. The expectation is that you will live to four score years and ten, that your life will unfold stage by stage according to some plan. The expectation is that there should be no suffering or minimal suffering and, furthermore, that there will be medicine or some other means whereby any suffering that does occur may be reduced or even nullified. The expectation is that we should have no pain, no suffering, and we hope that we may live long, be clothed, fed, sheltered...'

'...And the medical man will sort it out for us.'

'Precisely. Now, let's look at how life actually is. Do we actually know we will live to ninety?'

'No.'

'Do we know that we will live without suffering?'

'No.'

'Do we know that there will be somebody who will help us or put things right for us and take out our pain and suffering?'

'Yes, the homoeopath or the doctor?'

'No. The true scenario is we might die tomorrow and there is no power on earth that can stop that. There is suffering in life and we all know that. To wish it to be otherwise is a folly. There is suffering, there is death and we are suffering even now. More or less. That is reality and the other is a fantasy.'

'So why do we play to the fantasy then? How did it come about that medical men were seen to be curers?'

'Because they said they had means to alleviate suffering and we were so desperate to believe them we believed them. Suffering is the royal road, and the gateway through which we may pass in order to attain that greater freedom which allows us, not merely to be free of the particular symptoms of the disease, but more especially gives us that greater space within which to realise our fuller potential—the chance to realise that we are divine witnesses of the drama of life. More precisely, it allows us to give and receive more love in our lives.'

Misha's interest in medicine had begun when, at the age of six, he stayed in the house of the family doctor while his parents went on a three-week holiday abroad. Looking at medical textbooks to pass the time, the pictures of diseases profoundly affected him. They seemed to him to be somehow related to other human horrors—wars, gas chambers, mass exterminations. In his late teens, friends and family encouraged him to go into the field of healing, but he rejected the idea. He went off travelling for a year and a half, and on his return at the age of 21 threw himself into a career as a film-maker.

Cinema was not to be Misha's future, however. He fathered a child, 'a large and beautiful mistake', in his late twenties. The boy's mother was deeply disturbed by the responsibility and one day she just left them without so much as a forwarding address. Misha became a single parent overnight. He faced hardship and a new expanse of empty time. Rich prospects turned to poverty and the dole.

'The shock of events threw me out of the frame of being a film-maker where I had become quite identified with my profession. I had made a name, an association. It was a big package to do with ego fulfilment, with financial stability, paying a mortgage for Christ's sake, with my notion about myself. That was shattered. It was certainly emotional trauma but it was also practical. Needless to say it happened at about the twenty-eighth, twenty-ninth year of my life which is precisely when these things tend to happen. Saturn returns...'

To study was a relief, and Misha had a thirst for the wisdom of his teacher John Damonte. Homoeopathy was mysteriously successful. People came.

'John was not merely a teacher, he was the living embodiment of certain spiritual truths. In short, he was my guru, except that he would have laughed and split his sides if I had said that to him. He had the capacity of making complicated things seem simple, and guided me through the labyrinth of homoeopathy. He helped me to perceive what needs to be cured, to interpret the individual's own language of disease, so that I could really understand the story of their suffering. However after four years of this apprenticeship John died of heart failure, and I and his other students found other teachers to help us to continue our training. We made contact with another group of homoeopaths who had gathered around Thomas Maugham. When he also died, some members from the two study groups together founded the Society of Homoeopaths.'

The Society had grown into the largest national body, with a register of hundreds of practitioners, many of whom had studied with Misha. He had been dubbed 'the homoeopath's homoeopath'— but told me that he genuinely regarded this as a myth and was embarrassed at the currency it had gained. Still, from the early days he had been a big fish in a small pond, and thus was able to make the move from London to Devon and set up his school there in 1981.

But compared to his father's life Misha's was tame. His father Neushul was born in Bohemia, Czechoslovakia, and became the Supreme Soviet Artist as painter to Josef Stalin. He fled for his life when disgrace came his way, only to be hounded from his homeland by Hitler. Finally he had settled for Wales and died before his work became fashionable in the British art circles. His lugubrious paintings hung in the long stone hall at Yondercott, prized by his son who translated the family name to Norland to mark his new beginnings.

The boys in the garden had had enough. The promised game of hide-and-seek was long overdue. They did not want to hear their father point out grandfather's pictures again. Nor did they want to hear about 'the modern altars of ancient gods at which we pray'. Misha explained: 'One is Hecate, the goddess of rubbish dumps, and of those deep mine shafts in which we bury our toxic waste— an altar before whom the ecological movement daily worships. But each god and goddess has a natural counterpart: for instance, in in Hindu mythology the destroyer cohabits with the creator and maintainer, and the way of wisdom is to contemplate them both— and finally to realise that each is illusory.'

Lost in a wild corner beyond the formal gardens, I stumbled across a tipi set up for the summer. The creamy canvas gleamed and the clean pegs were well driven into the ground, but the plants around had not suffered a daily tread. It was a far cry from Talley.

We ended play-time for supper. For afters, there was a scramble to share the last eight chocolates between three boys and three grown-ups. The lads, dressed in skateboard pads, shunted each other around the back yard on anything available with wheels including my bicycle. We sat and chatted about Devon and talked about Brigitte's return to music teaching now the children were older. Misha and I left to walk the dog into the sunset. We passed the ducks in the kitchen orchard. They chased us off, squabbling when we turned backs on them.

When he described it, Misha's work had a different ring to it, compared with the more physical artisanship of Machynlleth. He was also well aware of a critique of the Green position. Sometimes he sided with the sceptics. 'The Green movement is involved in myth fulfilment,' he said. 'We have the biblical vision of the abomination, great fireballs rained down from heaven. This is the hole in the ozone layer. We have the flood, seen in the prognosis of the greenhouse effect. But whether we agree with this interpretation or not we can be in no doubt as to the fact that the story of Man, as described in the Bible, begins with the fall from Paradise in Genesis and ends with the destruction of the earth in the apocalypse.' The point was that the Greens had intentionally or otherwise hijacked traditional, archetypal myth for their own interpretation of events. It had been done before—was it so different this time around?

More worrying was the morality born of such invention, he said. Greens might come to believe that they were on the side of right and there was a tangible line between their version of good and bad which, once crossed, led to damnation. But wagging a finger from the moral high ground only alienated people. This would not win converts easily, it only cut off the campaigners.

We walked the fields and scrambled over ditches and under fences. I concentrated intensely on what Misha had to say. Health, he said, comes from inside the individual. It grows out in rings, like the rings of a tree, to the communities beyond. The philosophy of the American Indians summed up many very basic tenets. I was inspired to think of the many personalities I had met along the way. For so many the basic drive was that of home and community. The motivations were so powerful because one could recognise their rightness and good sense. Such values were not necessarily new or unproven—thrift, personal touch, small-scale business, social justice, land husbandry—and that was their power. With the Green philosophy of pluralism and diversity, you could group them together. And they made an immediate difference: small-scale business bred loyalty and honesty. You did not want to exploit your customers or your workers. You built relationships with the people in your community from the service you provided.

And, outside work, a person could be famous for their musician-ship or artistic ability. Their performance would be noticed and applauded by friends, not syphoned off and sterilised like mass pop culture. With a healthy village life, in contact with everyone from vicar to landowner, postmistress to teacher and children, there was the same chance to feel safe.

Misha and I were agreed on the spiral effect. If the first rings were in place, if you had a whole personality, your relationships with others were likely to flourish. There could be a way to bring vitality into the surroundings. And so on, inextricably. But where to start the healing?

Misha had seen so many people lacking strong roots to grasp. Perhaps they had lived in London for some years but had not begun to belong—they knew a few people who were in the same situation, but parents and siblings lived elsewhere. They were deeply unhappy with the prospects of loneliness, and their health suffered accordingly.

My experience was different. I had a respect for the communities that existed in London and reckoned that it would be possible to adapt to them if you were especially sensitive to their working-class heritage. These city villages perpetuated themselves through the local institutions, and shared their triumphs and hardships. In summers spent on the island of Lindisfarne I had also seen a village in relative isolation. I remembered in my own childhood I had been cosily insulated with cousins and family friends. Mothers' eyes beamed on me to encourage me to behave and ask the girls to dance. Emma's granny would mention my gallantry to my granny. They had played in the same hockey team at school; Emma's mother had taught my father to ride a bicycle during the war, though she could not ride one herself. Sally's mum was coming to lunch with mine next day. I was staying with Annabel and Jane.

I reckoned a healthy society came from sustained contact, from putting down roots, from nurturing them. With a close circle of friends, with a soul mate, you did not need to visit a therapist. You could talk out your problems.

Misha saw it differently, hinted at the privilege of my view and some naïvety in my assertions. He had known plenty of people who had never had a close family life. This created problems for them in many ways, and they needed a chance to talk through their anxieties.

We had walked full circle and crossed a waterlogged meadow by the garden wall; we were home. We agreed on the difficulties with either view: how to change lives devoid of any potential into a healthy, sustaining community. To convert people through loud-mouthed evangelism would be to deaden their idealism.

CHAPTER TWENTY

John Lane & The Dartington Hall Trust

T he weather had changed. It had been brewing for days. I had a
headache that nagged. The air had thickened overnight and
storm light made the creeper around my open window shine a vivid
green. A draught swept over me. My head throbbed. I diagnosed stress.

It would only be a few days days before I reached Totnes, and
my journey's end. I tried to weigh my new sympathies for the
Green movement objectively. I had started out ignorant of the
Green lifestyle, but after three months immersed in it I could not
easily distinguish between my old and new priorities. Why did I
care if I wasted another couple of gallons of drinking-quality water
flushing the loo? Did a small sprinkling of litter really matter? I
slept on rather than write my diary. I had doubts to contend with.
My optimism was tempered by Misha's criticism of Green
archetypal myth-making and the religious morality surrounding
it. Had I just been seduced by the latest cult in the first place?
What did it mean to believe that there was a supreme principle to
the Green ideology, namely, that if you harmed the earth you were
bad? How easy was it to prove harm had taken place and who was
responsible? Was my effort to alleviate that harm worthwhile, or
was I just kidding myself that the new lifestyle would be any less
draining on world resources? The writing was on the wall, but it
still depended on what you took it to mean.

These were philosophical reservations. Empirical evidence was

160

thin on the ground and the newspapers had started to pour scorn on the early claims of the environmentalists, with newly-collected data to counter the first assumptions. Doubts wormed their way into people's minds, because they wanted to believe, say, that their car fumes were harmless and that the future of the world did not really hinge on their little weaknesses for smooth, white loo paper instead of the rougher unbleached kind.

I still suspected that my small efforts were worthwhile. The lifestyle I had found restored dignity and self-esteem, even if others considered it a withdrawal from society into a deluded or pseudo self-sufficiency. But how did that relate to a commitment to change people? John Button's theory was that people had to want to change. Greens had to accept power in order to lead that change, as a prerequisite to progress, but that meant a duty to order other people around. Certain types of power were anathema: there was a fine line between forcing change and joining in the old structures to formulate new ones.

An earlier comment of Misha's came to me as I tossed and turned with these ideas. He had told me that such a fine line did not exist if one saw life in a transcendental way. Brig Oubridge had gone a long way from 'conventional' society and yet he found that he could not get away from that society's politics, laws and decisions. He was linked inextricably to the 'simple life', but only in comparison with the 'complex one' he had rejected. He lived a paradox, which made his simplicity very difficult to see simply. Fighters shot overhead just feet above his tipi. Their engines and technology were the fiercest physical embodiment of the complex life, and they also stood for philosophical conventions about the justice of killing. Brig wanted to fight that view because he wanted to deny it power. Misha was adamant that the fighter plane was as natural as the heather and bracken. The steel and glass and plastic were refined from the elements, and the plane had been commissioned as a result of certain human beliefs and values. Misha argued that there was a universal truth that every human shared the urges which had conceived of the engine of war, whether they deemed them necessary or not. We were all part-warmonger. The solution, said Misha, was to love everything, to be at peace with it in the soul and campaign for reform by example as well as action. The change was a state of mind.

Violence, aggression, powerlessness: I carried the full realm of responses around with me always, no matter how remote from the physical world, no matter how I tried to repress the anger and resentment. I had to contend with the big political questions of the human condition even as I lay and ate bilberries in the heather. There was no point in feeling bitter about violence; I had to try to persuade others it was not the way forward.

I woke late with all this tossing and turning, gulped down my muesli and set off for John Lane's house halfway across Devon. The wind splashed rain on to my face. I dug out my anorak and pedalled hard for Beaford. I would miss my appointment by an hour and John Lane did not sound like a man to keep waiting. He was chairman of the Dartington Estate Council, the latest in a line of experiments at the 800-acre trust just outside Totnes. John was one of six trustees with £30 million in cash and resources at their disposal to foster the new, the ecological, the holistic, the spiritual. They hankered after a complete renaissance of the countryside with responsible land husbandry, human-centred business and a nurturing of the imagination through spirituality and art. Totnes had benefited from the existence of nearby Dartington, to become some kind of capital for the Green movement. It loomed as my final goal.

My legs warmed up for Long Drag Hill out of Tiverton, but I felt the dreaded 'bonk' creep from my knees over my thighs. It addled my head, already spattered by the driven drizzle and steamed with condensation through the neck of my coat. My supply of the cherished antidote of Mars bars had run dry. I picked out a sticky wedge of nougat from my handlebar bag, the last iron ration I had bought at Harry Tuffin's Sunday Market on the Welsh border.

John Lane was cosily tucked in the bay window of his study at Kiverleigh Manor, an imposing late Victorian house facing South above Beaford village. I skidded to a halt on the gravel, bedraggled and feeling ropey, stowed my saddlebags in the porch under some old garden furniture, and knocked on the door. John's wife Truda let me in, somewhat astonished I had come at all, as the air looked so misty and murky. Mrs Lane disappeared quietly to the back of the house and left me to wait for John in the drawing room. I was surrounded by books waist-high in shelves, with china ornaments on top. Above, the walls were hung with pictures drawn by the mistress of the house. The room was bright and airy. I browsed and found Carlos Castenada's *The Teachings of Don Juan*, a book about South American Indian drug lore and its impact on a student from the United States, which Misha had recommended I model my book on.

I had read about three pages before John made his entrance. He had the bearing of a company director but was dressed in expensive casual clothes. A discreet purple jumper on a white-haired man in middle age did not jar with the total effect, given by the surroundings, of an understated, aristocratic and Bohemian elegance. We sat down to drink tea, which Truda spontaneously produced, and talked about the origins of the Dartington Hall Trust, informed by Leonard and Dorothy Elmhirst's vision and money. Elmhirst, the son of a Yorkshire squire, had volunteered for the YMCA during the First World War, then trained at Cornell University in agriculture. He had travelled to West Bengal and worked for Rabindranath Tagore, a philosopher and educationalist contemporary with Gandhi, and the first non-European to win a Nobel Prize. Tagore's estate had become a model centre for rural reconstruction.

In 1923 Elmhirst married an American heiress, and together they sought an estate to practise Tagore-inspired rural regeneration in a Western context. They bought the Dartington land and its beautiful stock of historic houses dating back to Medieval times. They had locked into an extremely well-developed estates tradition but brought also the modern American desire to experiment and test new ways.

There was a progressive school which lasted into the 1980s when rather over-publicised indiscretions and a pupil's death turned it into a focus for the backlash against liberalism that marked the early Thatcher years. An Arts college remained, and throughout the year festivals of music and art ran alongside conferences to foster spirit and mind. Small businesses run with values beyond the financial bottom line dotted the grounds.

Renewed vigour had returned in the nineties, spearheaded by, amongst others, John himself. As his watchwords he took those of Fritz Schumacher: 'Metaphysical reconstruction is the primary need for this generation.' He told me that a new college, called Schumacher College, was being established in place of the school to allow the awakened needs of this generation to explore ideas of wholism and spirituality. 'If we are really serious,' John told me, 'about the need for a new civilisation to come from the deeply blemished and soured one we live in, then in the long run the people will need a new kind of education.'

The college had invited a number of respected scholars to steer the student explorers, and the proposed list read like an international Who's Who of major global Green celebrity thinkers: James Lovelock, the man behind the Gaia hypothesis; Jonathon Porritt, ex-director of Friends of the Earth; Theodore Roszak, author of *The Making of a Counter-Culture*; Rupert Sheldrake, Cambridge don and author of *The Rebirth of Nature*; Edward Goldsmith, editor of *The Ecologist*. But everyone who had come to the college, whether student or resident, would share their insights and also the day-to-day support of massage, cooking and meditation. 'Universities since the Middle Ages have produced an élite education, all according to a set of Cartesian assumptions. If we are to go further to establish a new culture out of the ruins, we need a re-visioned university, a fully living environment.'

The ambitious plan was to create an active retreat for business-people as much as the converted, in an atmosphere balanced 'between the continuum of gentrified, countrified convention and extreme, rather outré things'. John had no illusions about the mishaps of the past, and focused on the future.

'We very much hope to involve middle management, perhaps with mini-sabbaticals for men and women in their forties and fifties to rethink their orientations. Business-people are extraordinarily enterprising because they have to be quick on their feet—whereas the university education system is like a dinosaur.'

Personally, John believed his emphasis on the neutrality and responsiblity of the new Dartington, together with its respectability, would encourage the 'ordinary person' in search of adventure. Milling flour or following permaculture were peripheral to the movements of society as a whole, he said. Schumacher College would prepare the ground for the right sort of people to take over if and when change was brought about by the ecological revolutions ahead. But this was only the beginning; he looked forward to a time when the whole Dartington estate was informed by ecological and spiritual values.

'We want to be ready to fill the power vacuum because I think people know that there are cataclysmic times ahead and that changes are creeping over us. People dream freely of an alternative. They are preparing for this in their subconscious. At the moment there are many people aware that they lead a schizophrenic existence. Though they spend their day living it, they are in concentration somewhere else. For instance, the banking system in the straight world is precarious. Our course of action must be to withdraw our energy from this type of system.'

For John, one philosophical answer could be found in art, which he sees in the broadest sense, as a way to instil meaning in work, relationships, even the laying of a table. He had moved to Devon in the first place to start an arts centre and worked there over five years before he started to work on the Dartington Hall Trust's Estate.

'In my book *The Living Tree*, I argued that the arts and the role of the aesthetic are marginalised in our society because the agenda is set by other disciplines. The arts have been profoundly corrupted by ideologues to make them bulwarks of archaic thoughts, to preserve and protect the preserves of ageing liberals. We need to smash the definitions and recast them. It has been done with health and attitudes to the body, but it's proven harder with the arts because Modernism is so deeply entrenched. I want art to resonate in the lives of everyone.'

Somewhere there would be a balance, perhaps metaphorical sunlight after the thunder. The words of Rudolf Bahro had been suggested as a motto for Schumacher College: 'When the forms of the old culture are dying, the new culture is created by a few people who are not afraid to be insecure.'

Outside, the rain and bluster had not abated and the clouds rushed by in mad mood. I trembled up hills and wolfed Mars bars and Double Deckers without flinching at the thought of the multinational enterprises that had made them, but my calves could not answer my call. I collapsed at Eggesford Station after a couple more hours' ride, and took the train.

CHAPTER TWENTY-ONE

Totnes, Green Mecca

Totnes looked rosy on a sunny day. Whitewashed old merchant houses in winding streets catered to the tourists' need for lardy cakes, English teas, the National Trust shop, pubs with twisted beams.

A Polish biochemistry professor and his doctor wife shared my bed and breakfast accommodation. Over the full English breakfast we talked about the tipis I had seen in Wales and the lifestyle the villagers led there. 'These people were lazy, yes? Such life is beyond reason,' the conservative doctor said.

I described to them the British utopian dream of country life. It was part of the British psyche to imagine living on a plot of your own and many fantasised that they could do so and be self-sufficient. An Englishman's home is his castle. The talk went on to landed gentry and second generation money and culminated in my trying to describe the hereditary composition of the House of Lords. 'The fact that by accident of birth people get into the House of Lords is to me beyond reason,' I said.

The Poles had come to England for the wedding of some friends in international academic circles. The celebrations were in a little Dorset village, and after the drive across Germany and Belgium, the professor and his wife had saved up for three weeks in Cornwall, the land of childhood fairy tales and King Arthur. None of the wedding guests had been farmers, which they had found puzzling—no farmers at a rural wedding. I suggested that they had witnessed the country cottage idyll in action. I explained that in Lancaster, where Rod Everett had sold his village house to professional types, prices were so inflated by the affluent classes that farm labourers were forced to live in the city. Every morning lawyers and doctors would journey into town, the farm workers out.

The commuter concept was beyond the middle European psyche. 'These are intelligent people, no?' the Polish doctor said. 'Why do they waste two hours every day in travel? Why do they want a garden? I too want a garden, but if I had one I would have a gardener also. For these intellectuals a garden is relaxation, you say?' The lady had a point and I was sick of all these Green dreams. Exasperation had taken over as I had waited three hours on Exeter station the night before, drenched in rain and exhausted by my tussle with wind and time. And I had been so looking forward to Totnes. It represented the culmination of my travels, the journey's end, the Mecca of the Alternative: Dartington, alternative health, small-is-beautiful businesses.

With stiff legs, I climbed Fore Street, full of luxury stores with olde worlde interiors. A shop called Paperworks, set back and no less timber-framed, stocked only recycled stationery. I decided to buy a new notepad to balance my negative attitude and reckoned that the shop assistant would have a fair idea of the local shrines that the pilgrims prayed at.

David Guest, the proprietor, recommended the Natural Health Centre above the National Trust Shop. He took down a pen-and-ink map of the streets and pointed out Arcturus bookshop, the home of a new float tank therapy. He arrowed Birdwood House Community Office, which had been set up to publicise Green Totnes and act as a base for initiatives that brought people together. There was Sacks, a food shop with a window full of advertising for the alternative scene, and the Willow Restaurant, which Misha had described as the Montmartre café of Totnes life. 'Rub shoulders with the verdurati.' The town could sustain not one but two separate rainbow cobblers, Green Shoes and Conker. Further uphill, Open Forum was a shop where Green groups could display their heady wares or conduct an open seminar for the browsers: permaculture and Friends of the Earth on Friday, market day, on other days therapy and healing.

David Guest suggested that I also went further afield in the South Hams triangle—Ashburton, Dartington, Totnes—sandwiched between the wilderness of Dartmoor and the wilds of the sea. He gave me a taste. Dartington Hall had an art school, a summer school and business park. Sharpham House, in trust from Maurice Ash, who had married an Elmhirst from Dartington, housed a

Buddhist community, biodynamic farm and rural retreat. Hazelwood House promised high culture and had satellite communities in Norfolk and Orkney.

'There's no movement,' said David. 'Not one nucleus. It's a mix. I moved here when my wife started at the Dartington art school. We found it was like London with the combination of people we could meet, unlike most towns of its size.' This was the encouragement I needed. I trimmed my sails, admitting to myself that the task of 'doing' Totnes would take a lifetime. This kind, smiling man had helped me pick up my faith. The sun had begun to twinkle through the showers of rain.

Andy Langford and George Sobol of the UK Permaculture Association headquarters in Dartington village were teaching a course in Yorkshire, but they had given me a few names of people to look up in their absence. Norman Duncan, the man behind the Community Office that administered the Totnes Green Pound exchange, and TILT, a local investment trust. Prem Ash was the boss of Conker Shoes. Serena Sobol, George's wife, was a homoeopath trained by Misha Norland.

The Open Forum at the top of the hill was shut when I passed by in mid-morning, looking for a phone box to call these insiders. The therapist booked for the day was nowhere to be seen, but the window was filled with elegantly simple meditation stools.

Prem was out; Norman was out; Serena was free for a hasty lunch. I bought a pair of braces in a second-hand shop and some snacks for Serena at the bakery. I ordered tea at Willow restaurant opposite and settled in to listen to what was going on. A soap opera of intrigue bubbled in my ears. I read the pamphlets picked from the rack to taste its public flavour. *Community* magazine attacked a critical BBC2 programme, called *The Totnes Alternative*, which had dented people's pride and dubbed the place 'barmy'; the magazine savaged the Beeb's 'sarcastic intent'. I took note, rather worried at the way my own sketches had turned out on the page. I looked at the advertisements. Four natural health centres were advertised in *Community* magazine, nine communities, twelve 'social organisations' including the Tibet Support Group and Compassion in World Farming, three meditiation and two yoga classes, and Narcotics Anonymous.

The local Men's Group leaflet sent tremors down my Anglo-Saxon spine. I remembered John Lane telling me that the Elmhirsts' American have-a-go spirit infected the area. Ideas thought freaky elsewhere thrived here. The leaflet outlined the idea for a men's walk. Four groups of men would set off to walk from the four compass points to meet in the middle of Dartmoor and celebrate. I read on, intrigued. The father figure of the movement was an American poet called Robert Bly, 'strong precisely because he is willing to make himself vulnerable'. A Men's Group story by David Guy, reprinted from the *North Carolina Independent* satisfied my curiousity about what men needed to group together for:

'We talked often about our fathers, a subject that never fails to summon intense interest. At the end of the evening we gather in a circle and put our arms around each other. We smile and start to laugh. There is probably some embarrassment in those smiles—men don't do this—but there is also a sense of relief, and of unabashed delight. This is the way things should be... a place where I was entirely who I was, where other men were being who they were and that being ourselves gave us a power that no one of us could have had himself.'

I read three other flyers and sipped organic tea. Sharpham North Community, comprising nine Buddhists from across the globe, organised meditation on themes like '365 days in a mud hut', 'Seeking the spirit behind the matter', 'Subtle energies: A Tibetan Buddhist demystification.' Also on the Sharpham estate, the Barn promised mental retreat and hard labour on the land: three and a half hours of one, five of the other, every day.

The Devon School of Shiatsu and Macrobiotics advertised the life-changing philosophy of George Ohsawa and finger-pressure massage techniques. George updated oriental traditions in combined health, self-awareness and spirituality. He considered diet an imperative, although in a note from the school's director Oliver Cowmeadow, visitors to the Devon School were asked to bring their own light lunch.

I was beginning to wonder how many ill people there were in Totnes to support so many therapists. I had never encountered so many different cults based around the teachings of gurus.

'You may not be ill yet, in the conventional sense of having symptoms,' Sebrena Sobol reminded me over lunch in Dartington village, ten minutes ride from town. 'But there may be underlying problems.' Serena did find odd the Totnes emphasis on obscure therapy, from crystal healing to a fairy-tale fringe. The therapies were mostly based on real results. Yes, Dartington Hall Trust had long ago set the tone, she agreed. For example, they had harboured a German ballet company throughout the Second World War to save its Jewish dancers. It was also a conducive place to live, with moors and sea nearby, and also the big cities of Plymouth and Exeter. That 'special blend' that had created the modern Totnes had taken off in the eighties.

Homoeopathy was part of the blend. Six years earlier Serena's doctor had given her a homoeopathic remedy to calm an abscess in her tooth. She had felt better immediately and recognised the benefit of working with the body to encourage its healing reflex, instead of attacking it outright with drugs. The body had already done its best to discharge the disease, it just needed some extra energy. So she had decided to train as a homoeopath herself, and enrolled at Misha Norland's School of Homoeopathy. Serena likened her homoeopathic craft, translating the Greek root as 'similar suffering', to the martial art of Tai Chi. In Tai Chi one trains to turn out of a blow and to go with its force, so as to minimise the overall shock. If you stood your ground and fielded the hit you risked being pushed off-balance altogether.

The simplicity of truth influenced Serena's decision to study but she also felt safe that homoeopathy was driven by some hard principles which could be understood in the 'straight world' context. An atomic physicist could understand that lost energy needed a boost from outside: the question and answers were empirical. So, to an extent were homoeopathy's. 'I saw the practical principles crop up in other therapies, in nature and in art,' Serena asserted. 'I could see they were true. The doctor who treated my tooth abscess was not just a nice man.'

She and I sat and ate chopped tomato and olives in the kitchen as her son Max whizzed up his second chocolate milk shake in the liquidiser. 'When he's on holiday, he doesn't realise I am still at work,' she said, with a sigh for the bliss of childhood.

The phones never stopped ringing—one in the hall, her office, and one in the kitchen, the UK Permaculture Association's headquarters. Patients asked for snippets of advice, and the Permaculture Association had been overrun since Channel Four featured Bill Mollison on its *Visionaries* series. Hundreds of people had called to enrol for courses and fifty-four Third World countries had asked for help the permaculture way. At the same time there were so few qualified teachers that an Australian had flown in to oversee the latest course which George Sobol had organised in the North of England.

Homoeopathy was somewhat better established, so Serena could plug into a circuit of institutions and support groups. She unwound with a supervision group of other therapists who discussed how clients took their toll on the practitioner. The group also shared their research discoveries, to compare information and key remedies. As a group they subscribed to Homoeonet, a computer channel which bridged the Atlantic with case and remedy notes.

Serena had joined a clinic group, not part of the Totnes Natural

Health Centre, to put her work on a more business-like footing. Like a doctor's surgery, someone was always on call for acute problems—a pregnancy pang, severe bleeding or screaming sick kids. And Serena could get cover for her patients from other members of the group when she went on holiday. This helped her to separate work and home life.

Treatment cost £30-£50 for the first visit, £20-£30 each hour thereafter. But for every hour with someone, Serena spent two or three considering the case and most often she would prescribe a remedy much later.

She was just back from a week-long festival, a small-scale Glastonbury, where first aid was all homoeopathic. Her team had seen 300 patients. Some had caught camp 'flu, some had had sudden sickness, most had had a long-term illness which had flared up with the stress of a change from what Roland Chaplain had described as 'the security of one's discomfort'. A Kenyan performer had had a bout of aches Serena put down to severe homesickness. 'Is there any other way you feel under the weather?' she had asked him.

'Yes. In Kenya the water tastes fresh. Here I cannot drink it unless it is bottled.'

Everything is relative.

I dashed back to meet Norman Duncan at the Community Office, and locked my bike against a pile of waste cardboard ready for collection on the street outside. In the window of Birdwood House hung a child's notice: "We are helping to mack a communati and saving the invironment."

The idea that a community could be made and the environment saved was Norman's. His centre would be a good beginning as a drop-in meeting place, a library of local resources, a foundation for an ecological town. He applied for an enterprise allowance and was offered a high street base by the Birdwood Trustees, caretakers of a house by the market given over to benefit local people.

Julie and Ian were on Employment Training while they studied permaculture and counselling. On the wall, a chart explained their mission, from individual health to recycling and clean air.

'Totnes is a place where people have all sorts of ideas about how the universe works,' Julie told me. 'Yes, there are cliques,' she went on when I questioned her. And navel-gazing? Totnes had a good share of experimenters less grounded in the world of proven fact than were responsible homoeopaths. Yes, there was exclusivity in Totnes and this led to the more cerebral of pursuits. The town had never had any large working-class contingent. 'Totnes is a safe haven for people to take the opportunity to sort out issues for themselves. There are lots of genuine, like-minded people so there is security to learn and experiment in new ways to deal with life.'

At the Natural Health Centre, Julie told me, I could learn Alexander Technique, a complete form of body deportment and harmonious thought developed to optimise health. It was a process built up by an Australian actor, which I took to mean that it was based on approachable Western tenets. For the more extravagant tastes I was recommended Karuna, Buddhist psychotherapy, which was taught at Foxhole on the Dartington Estate.

The Community Office business was on a different level. Link Age, a scheme to bring generations together, had won an Institute of Social Inventions award in 1990. The project was conceived by Norman because he did not want his daughter to grow up without contact with her grandparent generation. The South Hams area was full of parents too far away from their elders for regular visits and the coast was also full of retired folk. Link Age married the generations like a dating agency, as part of the grand plan to 'mack' communication and community. Julie reported that take-up for Link Age was a touch disappointing because they had not got the right balance of personal contact needed to build such intimate bonds. The hitch-hiking register, another initiative, also needed more time to develop, until the habits of drivers could be transformed into thinking about filling their spare seats.

To my disappointment, the Green pound system of barter, which Pat Fleming had described and David Stephens had tried to emulate on a national level, had also been shelved for the time being. All were agreed that the Green pound was a laudable idea. A Green bank effectively created its own local currency for transactions. The Green pound acted as a step beyond person-to-person barter but still kept traders beyond the taint of national cash circles. Instead of 'I mend your car, you give me massages worth the value', the deal would be weighed in Green pounds. So the masseur didn't have to give massages to the mechanic for the rest of his life but earned credits from the baker and candlestick maker who were also on the register. These could then be paid out like ordinary pounds. 'It is a way of exchanging as well as a forum for people to explore ideas of giving and receiving,' Julie said. 'But we had problems keeping it going. It's about ready for an energetic person to take it on again.'

Rather than hammering at these projects to create a going concern, it appeared that the scatter-gun technique was being used. With a load of shots at the target, one might hit the mark. Norman's latest project TILT—Totnes Investment in Local Trading—was conceived as a way to hold investors' money inside the local area, along the lines of those original building and friendly societies. The first loan had gone to an artist to back her first exhibition. 'The "interest" is not only on your money but from a personal share in meetings with borrowers,' Julie said. 'This can demystify the world of finance. It is not an alienating experience, therefore it is an education opportunity.'

I finally tracked down Prem Ash at the Conker Shoes production shed behind the shop. 'But none of us here consider Conker remotely Green,' she said and rushed to give me a leaflet about a worthier project—TILT.

In the environmental sense, perhaps Conker was grey with its leather and dyes and rubber. But the business was Green in its care for the customers and workers as individuals while maintaining its viability. Shoes were affordable and unique to each person's requirements but this did not necessitate an army of cobbler slaves.

Conker had been set up by Andy Langford, who had since transferred his energies to permaculture design. He had began his working life at Clarkes, the shoe manufacturer, trying to improve split-second timing along the production lines but gradually he realised that the factory removed people's dignity. And when he had tried to buy his daughter some shoes one winter, he had found that none of the mass-produced boots fitted either. In a fit of pique, he had said to his wife, Verity, 'We can do better than this.' And he had.

'The beauty of the project was that anyone could get involved and become useful within three hours,' Andy had told me. 'They could help out in the Conker processes for a short stint on their way through life's journey.'

The shoes, bubble-shaped in the company's distinctive rainbow colours, came in twenty-seven different colours and four different fits. The customer either drew around their foot and sent in the picture or tried on a comparable pair at the shop and the Conker workers could compensate for any unusual contours. Any part of the finished product could be replaced at a later date. One pair had just returned after ten years but usually the shoes lasted two years before they came back for a refit.

Prem, the boss, was dressed like the other cobblers in a smudged apron to grind down new soles and stitch the leather. Every day they concentrated on a different part of the process so that nothing became too repetitive, but they managed to speed through batches at a time. Business was booming, with 100 pairs made each week. This was enough to support thirteen workers on between £3 and £3.50 an hour, depending on experience. And the business was so organised that no one worked full-time at

Conker, thus following Andy's ideal of a short stint on life's journey. The maximum was four days a week, to make free time for other interests.

Turnover had risen to £150,000 a year, Prem told me, as she stood in the cottage factory. At the back of the long narrow room, three young men crafted final shapes on spinning lathes and stacked various pairs of shoes. They turned to stack them on a shelved trolley taller than the woman who wheeled it away. She logged each shoe with a tick in her book, a ticket inside and a code on the shoe's tongue. At strategic corners around the benches other trolleys stood ready for the next step. Colourful off-cuts and laces hung from every beam in the room.

As Prem was boss she could take a break for some PR with me. She told me the company had outgrown the cottage and had had to break up production, with a warehouse assembly plant in Dartington. She was sorry that the site would be split, but had reconciled herself to the growth. But the company had not been offered a building on the Dartington estate. 'If only we could get our hands on some of their resources,' she said. 'But they are very precious about them up at the Hall.' She voiced a tetchiness felt in town by the many who never shared the Dartington Estate wealth. Dartington preached its gospel and could afford to sprinkle favourites with gold, while others had to just soldier on.

Andy Langford's original concept for the Conker business was to teach people the simple craft of making simple shoes so that they could then start their own shops across the country. His partners worried that unmasking their trade would lead to competition and their extinction, whereas Andy just hoped to transform the shoe business with his human-scale alternative. A different view had informed the wisdom of the next generation, it appeared. They had not followed Andy's vision of vertical integration with an organic tannery and other peripheral businesses. They wanted an expanded production process.

I asked Prem how she gauged the atmosphere of Totnes. She told me: 'The town is full of people who are doing interesting things with little reward or stability. It's full of people who think they are doing something valuable and aren't...'

'And,' interrupted the woman wheeling the shoe trolley, 'people do a lot of talking.'

I had outstayed my welcome. Talking was not acting. I had a train to catch and a book to write. I walked down the narrow street balancing the heavily laden KP Crisps, which had tortured me for so long. The shops were shutting up for the night and the temperature beginning to chill. I was homeward bound on the next train to London. The summer was over.

174

The Natural Health Clinic, which everyone had recommended, was closed for the day. I wasn't really bothered, for some reason. All this healing and 'channelling' and meditation was not really for me. I might have been ill or dying, but ignorance was bliss.

For a last thrill I sauntered into Arcturus Bookshop. I browsed through the noticeboard of business cards at arms' length: tarot, Gestalt, reiki, crystal healing, hypnotherapy, Bach flower remedies, focusing. I saw no reason to give these therapists the key to my confidence or my bank account. On the floor above the bookshop's crystal section was a particularly zany remedy to the excesses of the twentieth century: Oasis Float Centre, run by Francis Checkley, a smooth, straight talker in his forties dressed like a swimming pool attendant. He had been in the floating business for two months and the centre, with only one chamber, was already on target for a profit.

Francis told me that he had offered a twenty per cent discount, one free float for every five, to the other Totnes healers if they referred their clients to him, and yet less than ten people had been sent. He could not work it out. Here was a chance to experiment with a different medium for meditation or therapy or healing. Did they fear competition? A dilution of their methods?

What was floating anyway? Francis took me to see the flotation chamber. Clients stripped in a bathroom and showered down. Then they went through to the chamber, and lay down alone in ten inches of water, to float. Francis opened it up. The smell was off-putting. Some 700 pounds of Epsom Salts were dissolved in the water. They were being filtered as a client had just left. 'The skin descales, you know,' said Francis. He told me floating would be a breakthrough for me. I could see that lying outstretched in complete silence, in complete darkness could have its advantages for getting rid of the world's stresses and strains without having to bare my soul to a third party. Being suspended without sensation would be like death, but for the smell of the Epsom Salts and my descaling skin.

I realised, belatedly, that I had asked many people about their lifestyles and Green principles, I had trodden a path of minimal environmental harm for the future of my children's children's children, but I had not asked what state of mind these people took to the grave. The nearest I had got was discussing pensions at Laurieston Hall and papier maché coffins with Andy Moore, in Bristol. I settled back in my seat on the train to ponder this as we pulled past a sign on the platform which read: 'Totnes Green Zone'. There was something about sitting on a train, knowing you were homeward, that surpassed any of the claims of a float tank.

On the train home, I settled back in my seat and thought about death. Before I go, I must book one of Andy Moore's papier maché coffins, and be planted in an orchard. But there was something about being homeward bound that turned my thoughts from this, and from any reflections on the pros and cons of the Green Life. A conscience is for the waking hours—the rolling of the train lulled me into sleep.